Building Research: Design, Construction and Technologies

Series Editor
Bárbara Rangel, Civil Engineering, University of Porto, Porto, Portugal

Building Research: Design, Construction and Technologies brings together knowledge from civil engineering and architecture to make an interdisciplinary analysis of a recognized building/project or a building construction research. In each volume, the topic is a building or an aspect of a building that catalyzes the contributions of invited authors in their fields of specialization, with professional and academic backgrounds. To make the bridge between the scientific research and the construction site problems a parallel reading of the working development and the technological issues raised by that construction problem are presented. Authors, architects and engineers are interviewed and analyze the different projects in distinct stages, from concept to construction drawings, following the development of the building's design and construction process. The series treats topics such as building technology, construction management, acoustics, maintenance, prefabrication, amongst others. As a complementary objective, authors with different backgrounds: Engineers and Architects, Researchers and Designers are invited to to contribute to the understanding of specific buildings through the analysis of those issues that influenced the development of the design or that appeared during the construction or the facility management phases.

More information about this series at http://www.springer.com/series/13572

Vítor Abrantes • Bárbara Rangel
José Manuel Amorim Faria
Editors

The Pre-Fabrication of Building Facades

Springer

Editors
Vítor Abrantes
Civil Engineering
University of Porto
Porto, Portugal

José Manuel Amorim Faria
Civil Engineering
University of Porto
Porto, Portugal

Bárbara Rangel
Civil Engineering
University of Porto
Porto, Portugal

This is an updated translation of the original first edition published by Gequaltec.

Building Research: Design, Construction and Technologies
ISBN 978-3-319-22694-1 ISBN 978-3-319-22695-8 (eBook)
DOI 10.1007/978-3-319-22695-8

Library of Congress Control Number: 2016937330

Springer Cham Heidelberg New York Dordrecht London

Printed on acid-free paper

Springer International Publishing AG Switzerland is part of Springer Science+Business Media (www.springer.com)

Editorial

This book represents the first of a series designated *Building Research: Design, Construction and Technologies*. The series aims to give a positive feedback to a Springer's invitation to organize a book series on building construction using construction as a common field of work of architects and engineers and putting together research and practice.

As the group has been doing in its publications at the Faculty of Engineering of University of Porto, this series therefore proposes to put together the knowledge and complementarity of different areas involved in construction, engineering and architecture, through two large binomials, RESEARCH AND EDUCATION / PROFESSIONAL PRACTICE. For this series, the group proposes to cross the construction design and scientific research in the analysis of a thematic issue related to building research. To make the bridge between the scientific research and the construction site problems, this books series will propose a parallel reading of the working development and the technological issues raised by a construction problem. The analyses of one or two case studies, recognized projects in the international panorama, suggest topics related to the building research, such as building technology, construction management, acoustics, maintenance, prefabrication, etc. These are the themes that authors with different backgrounds (engineers and architects, researchers and designers) are invited to contribute to the understanding of the problem.

For the first issue of *Building Facades Prefabrication*, we selected two buildings: one in New York, by Rafael Moneo; and the other in Porto, by Eduardo Souto Moura, analyzed in the first issues *of Cadernos d'Obra*, a scientific journal on construction we edit at the Faculty of Engineering of the University of Porto. The most recent building of Columbia University in New York, the Northwest Science Building, a four-hand design by Rafael Moneo and Dan Brodkin of Ove Arup, designed in 2005 and opened in 2010, and the Burgo Tower in Oporto, a building that brings a new perspective to the use of prefabrication technologies with local traditional construction systems.

Besides its importance from an architectural and urban point of view, these two buildings suggest interesting topics that are present in current building research, such as the construction of facades in high rising buildings, prefabricated facades, construction safety issues related to the development of this solution, etc. The scientific analysis is accomplished not only by presenting the architectural aspects but also to connect those aspects with management, technology and building processes in adequate depth. Alongside the case studies, specific articles on engineer and architecture explain the critic issues of these buildings, from the analysis of specific technological problems of such buildings, to the critical reading of the relationship between design and technology, and to the testimony of the designers involved in the projects.

This first issue reflects not only how closely the professional and academic areas of the construction sectors are connected but also the urgency of this connection to fulfil the current requirements of construction science and its two main fields, architecture and engineering.

For an architect, designing prefabrication could be an outlandish but stimulating challenge, because it implies particular concepts and project methodologies. Over modern architectural history, some architects explored the potentiality of the optimization of this constructive system to innovate in design conception. Wright explored the modular capacity of the brick to design the Usonian evolutionary houses. Corbusier found the modern house formula in the Dominó house based on the prefabrication of the concrete structure for the house. Mies van der Rohe upscaled the *Mechano* system concept to design with steel when he arrived in the USA. Richard Rogers learned from the car industry the construction systems to increment a house. In all of these projects, it is possible to identify a common design strategy, using the modularity and repletion of the industrial production into the building construction methodologies.

The current difficulties in the construction industry have confirmed the urgent need for interdisciplinary action between all the design disciplines, including all the different engineering branches and architecture. The requirement for high levels of building efficiency and the optimization of the building process are making increasing demands on the accuracy of designs. The project is no longer a sum of contributions, but a design methodology that combines the answers to all the different requisites of the building, an INTEGRATED DESIGN PROJECT. This multidisciplinary approach to design problems is only possible if it is present in the design process from the outset.

These days, a building is required to be as efficient as any domestic appliance, but there is probably not a single machine produced by human that involves so many systems and people. The performance efficiency demanded to buildings in service by the different stresses to which they are subjected has been imposing in recent decades an increasing rigour in the definition of the various projects, particularly in the degree of scientific and technical depth in the solution presented. To reduce the construction time and anticipate this building performance, greater specifications are expected from the various disciplines involved in the design for the constructive solutions they propose for each subsystem. It is the perfect articulation of all these subsystems – structural, hydraulic, acoustic, electrical and others that will establish the perfect functioning of the building as an INTEGRATED SYSTEM. Architecture, thermal insulation, acoustics, structures and hydraulics inevitably come together to design the solution because the same building elements often belong to different subsystems. The same wall might be the structural support for a slab, an acoustic barrier in a room or a thermal barrier to the outside.

From the initial sketches to the specifications of the various materials, the design cannot definitely be a sequence of responses, but the result of a complex algorithm of the different responses of the various disciplines, the INTEGRATED DESIGN. The resolution of this algorithm, the DESIGN OF THE CONSTRUCTION, lies in the common ground that underlies all the disciplines. In determining the dimensions, materials and construction solutions, each member of the team must identify the solution that meets the requirements of his or her discipline. The optimal solution corresponds to the weighting of the responses from each of them. In designing a room of 200 m^2, for example, the architect cannot design the openings to the outside without knowing what the ideal relationship is between natural and artificial lightings. The architect cannot define the shape without the acoustician determining the ideal volume for the interior space or else he is not able to design the finishing materials without knowing the desired degree of reverberation. Also, the architect cannot scale the form without knowing the dimensions of the horizontal and vertical structural components needed to achieve a span of 15 or 20 m. The final solution for the form, spans envelope, for example, is therefore the result of the articulation of all these decisions. It is an algorithm that unites the optimal values of each discipline to find the suitable dimensions of the various components of the architectural form. Prefabrication systems help to simplify this equation, part of the solution is a non-variable parcel, which gives the architectural design more freedom to invest in the design phases. The comfort performance is assured by the prefabricated components, which gives more time to the design team to invest in the architectural and design issues.

Some ideas arise from this book concerning the more important advantages of prefabrication. In the following paragraphs, as an "appetizer", we develop three of them which are as follows:

- Prefabrication widens the space for industrialization in building construction narrowing the distance between this field and its cousin industrial engineering.
- It constitutes a very powerful tool for architects if correctly used.
- It constitutes one of the easiest ways to introduce innovation in building construction.

Prefabrication and Industrialization

Since the beginning of civilization, humans have always tried to perform their basic tasks in a more or less repetitive way in order to try to reduce the amount of effort and energy needed to do them.

It is possible to find examples in many areas, from agriculture to the confection of meals or the production of tools.

The pursuit of increasing the productivity (reduction of the amount of time spent doing a unity of a certain task), or, expressing it in another way, reducing the effort spent to achieve the same goals, has always represented therefore a major concern of humankind.

Therefore, at the construction level, industrialization has been assumed as a fundamental solution to accomplish this essential basic goal.

Prefabrication represents one of the main fields of construction industrialization. Currently, prefabrication is defined as "a set of construction techniques that are based on the production of construction elements outside of their final places of definitive setting, on site or in an external production unit, which are afterwards connected and assembled on site".

The industrialization of construction may solve many problems such as to reduce the time needed to build (using prefabrication, many times we say to erect in opposition to build on site…) and to avoid difficult weather conditions.

Concerning cost, the situation may be a little different because prefabrication demands in general huge investments, and therefore, local site construction costs with labour and materials are substituted with costs with industrial facilities and equipments and the final trade-off of costs depends on a lot of "market issues".

In macroeconomical terms, industrializing construction is very important for the nations and the economical zones because it facilitates exports and increases quality because quality control procedures may be applied in a better and more efficient way.

Prefabrication and Architecture

Prefabrication implies a certain discipline on the conception related with geometrical organization and repetition which also, since always, represents a barrier to its penetration on the fields of higher importance to architects.

Therefore, in general, architects have always considered prefabrication as an enemy to creation liberty, a prison which they didn't want to submit to. For an architect, however, prefabrication may constitute a motivation and at the same time a paradigmatic inspiration of the creative process.

However, prefabrication may be a very powerful tool for architects if correctly used. Using prefabrication implies to be able to use in a wise way the "tools" of prefabrication design which are mainly modular coordination, control of construction tolerances and correct and complete design of joints. If these three issues are well controlled by the architect, using a specific design methodology gives an incredible strength to architects to consolidate and explode simple ideas.

Prefabrication design has a lot to do with building integration. Nevertheless, connection and integration arise from the "superposition" of volumetric very innovative three-dimensional modules, where each module is very carefully designed, planned and fabricated in an industrial plant. Alternatively, the design may be produced in a planar mode where pavements, walls, ceilings and roof tops are designed as a succession of parts put together with "joints", respecting a modular basis and managing to solve all the construction tolerances issues. Architects try to learn/copy other industrial activities such as the automobile industry, the "all-the-time" leading industry in innovation, prefabrication, automation and industrialization.

Prefabrication and Innovation

Prefabrication constitutes one of the easiest ways to introduce innovation in building construction. Innovation comes from very different origins such as academic research, internal research and development in companies, patents and original ideas.

Construction began as an artisanal activity and has recently developed into a more industrialized and mechanized industry, profiting initially from the eighteenth- and nineteenth-century industrialization (although with a large delay in relation to other industries...) and from the more recent developments in all fields of science with an emphasis on materials science and CAD/CAM machines.

Prefabrication transfers production from each specific site to more or less developed and skilled industrial facilities where industrial tools may be better used. The production of construction components and "global construction systems and sub-systems" in a factory reduces the risk of mistakes, facilitates compliance with technical specifications and enhances productivity. In addition, it allows producing "high-tech" components, planar or three-dimensional, using modern industrial tools and machines.

Thus, one may consider that in these facilities translating research into practice becomes easier. Maybe this is why one might consider prefabrication as the best way to introduce innovation in construction.

Enjoy finding in this book specific ideas and solutions used in the interviews, the articles and the two real-case examples that illustrate these views!!

Contents

Part I

Burgo Tower in Porto

Abstraction and Traditional Materials in a High-Tech Façade

Tectonic Design

Conversation with Eduardo Souto de Moura on Burgo Tower about Engineer and Architecture

Bárbara Rangel, José Manuel Amorim Faria, and João Poças Martins

In 2008, for the first edition of CdO, we interviewed Eduardo Souto de Moura, to understand the development of the project of the BURGO tower in Porto, a building that crosses contemporary technologies of building facades with traditional Portuguese materials. The concept for the façade was a challenge for the architects and engineers. Traditional Portuguese granitic stone was used, which is heavy therefore difficult to manipulate during the construction. This building would certainly not be possible without a strong complicity between architecture and engineering since the beginning of the design until the construction of the façade.

To understand the particularities of the development of the project, the preparation of the building works and the construction of the building we explored four main themes in these interview. To understand the project methodology, we explored the relationship between ARCHITECTURE and ENGINEERING during the design and the construction phases. To explore the design concepts adopted, we tried to discover how the CONSTRUCTION RESEARCH influenced the architectural design development to define the building, the TECTONIC OBJECT. Finally we tried to discover how the CONSTRUCTION DESIGN determines the whole project in Eduardo Souto de Moura's way of work.

The façade is nowadays understood as a skin, particularly in tall buildings. This concept has overpassed architecture and become in some cases the stakeholder requisite. The façade is becoming an independent system of the building. Is this transforming the design methodology for architects and engineers?

Today's architecture is divided into two different kinds of interventions: in the first kind, clients and architects require images that are materialized using the available technology. Engineers often have to suffer to cope with these kinds of images. And using today's technology, steel structures, cantilevers, the T's, the X's and so on, they actually manage. Then there is another kind of architecture, that doesn't show off as much, although architecture always has been a show, there is nothing new about that. Popes have always wanted a show, so did kings. This other kind of architecture doesn't have these icons and doesn't build monuments, it is dedicated to residential buildings, buildings that are more anonymous, a kind of architecture that portrays its time, its own culture. As Mies van der Rhoe said, we must make architecture with the possibilities of our time. In those cases, I still believe (I am a dinosaur) that there is a close relationship between material, language and building system. This relationship can then be changed, concealed, inverted, but it is a starting point. Without that, anything goes… It's not bad, but it is very tiresome. One would have to imagine a concrete building wrapped in plastic or a plastic building wrapped in carbon… So the starting point is always that one, that reasonable base point, adequate material, adequate building system and meaningful language.

Sometimes that language falls short of what we intend, of insinuations we want to make and must be changed. That's where the building skins come in. But that isn't a kind of architecture that has interested me so far… I feel that resistance is becoming more difficult. That other kind of architecture is imposing itself through an attack, both by private clients and by image-based buildings. Public administration wants iconic buildings that leave a mark on the city and

B. Rangel (✉) • J.M.A. Faria
Faculty of Engineer, Oporto University, Porto, Portugal
e-mail: brangel@fe.up.pt; jmfaria@fe.up.pt

J.P. Martins
Civil Engineering, University of Porto, Porto, Portugal

V. Abrantes et al. (eds.), *The Pre-Fabrication of Building Facades*, Building Research: Design, Construction and Technologies,
DOI 10.1007/978-3-319-22695-8_1

change it. Politicians want to leave a mark, but things were always like that!

Do you think then, that the way things are going, that the value of an image and the value of a brand are becoming decisive?

I think they are undermining everything at the moment. Books' covers keep looking nicer and more attractive regardless of their content. Architecture must have a certain atmosphere… It starts at the tender process, before cross sections, elevations and floor plans are developed, appealing 3D renders are already available. I notice that, not because I lose many bids (almost all of them actually) but because the same ones keep winning them, the ones that develop the most appealing images. And it isn't by chance that when these architects build, they are the least interesting in my opinion. Nowadays there is a group of architects who follow a more, say, conservative approach that need to build in order to show that they are good and the ones who can draw well who need to build so they can say they're good. It's a little contradictory.

They follow opposite paths. In that sense, when we speak about images and the commercial sense of images, so to speak…

That is not an architectural problem. It is a general problem. I mean, today's culture is built upon images, content does not matter in any product, either in advertising, in literature, in painting… When a painting is abstract, its content disappears. It is "oh, so nice"! Architecture is also "so nice"! It is all upside down.

Architecture is not the manifestation of an individual who decides as he or she pleases. There are too many "artists"… Architecture is a social act.

People have forgotten that architecture and engineering, which are almost the same thing, occupy places. Geography belongs to everyone, so I have no right to go somewhere and impose a formula, or a stadium, or a housing project or anything else just because it is *my* site. It *is* my site but it is not my image, it is everyone's. That's why architecture is a social act.

Things must be thought through so that people will later adhere to the project. In the Braga Stadium, for instance, people liked it. People from Braga call me an engineer and they take pictures with me. This means they adhered. Today, the stadium is (and I am not vain!) an object of affection for Braga's people. People go on day-trips there.

Are we driving again to the concept of façadism?

It is not just façadism. Here we have to speak about baroque. Baroque was façadism. The problem is that baroque and baroque's façadism had the Counter-Reformation as a background. There was an ideology that intended to show that Christianity hadn't been hurt by the Lutheran reform. So they invested in a new language. In the popes' concilium, new religious orders were created (like the Jesuits) and new architectural styles were developed, which resulted in well known façades such as the "Je Jesu". Nowadays we have the images without everything else in the background. When a twisted building appears or they say that the floors go around, I don't see Jesuits, nor popes, nor the Counter-Reformation. Why should someone go to sleep facing West and wake up facing East?! I think that this support is not there. There is "more aesthetics and less ethics", "form over function", when up until now, there was always some balance between form and function, one supported the other. Every style (for example, classical and so on) would show different images but would have a supporting background. Nowadays one impresses through difference, and that's enough.

Construction is the common discipline between architects and engineers. The definition of the constructive system that will define the building is a consequence of the combination and integration of the disciplines involved: architecture, structure, fire safety, among others. The concept defined for each one of these disciplines is filtered by a set of requisites and regulations that will beacon the solution achieved. These regulations are supported by scientific research in construction.

Is it important for an architect's notion of construction to understand the comfort engineering and not only about structural engineering?

If I were to start a school, I would teach only three subjects in six years: Drawing, Building Construction and History. Throughout the 20th century (and there are some traces about this in the 19th century), sciences moved outside their original boundaries. Some branches such as chemistry, physics or biology, follow methods and results that are then lent through analogy to other kinds of research, to other subjects. In order for History to advance, anthropology is needed, because if anthropology can study the behavior of an Amazonian Indian or an Australian native and understand what life was like in the past, it can understand many ancient social structures from the study of western societies. Therefore, History needs Anthropology, Anthropology needs Sociology, Chemistry cannot advance without Physic's studies about DNA and neither can Biology. The exact same thing happens in architecture. If the artist-architect has a sketchpad and he draws shapes, it can be very interesting as an egocentric activity, but in order for them to be useful and materialized, he needs information. Nowadays, an architectural object must meet a wide set of requirements. The object must be correctly materialized, cheap (there is no money!), comfortable (people demand comfort) and sustainable. You can't have a glass pane facing South with the AC turned to its maximum! The structure must be reasonable, I won't even say good.

So right now Architecture really needs Engineering and Engineering needs Architecture. I think that subject is outdated because there is no architect in the world who doesn't work with great engineers from day one. In magazines you can see great designs by Koolhas, Herzog, always with the same engineers from ARUP. The same thing happens in Portugal. We need an engineer from day one because our ideas must be filtered. Today I cannot start a new project or go anywhere without an engineer, even just to take a look at a site. First of all, because it's pleasant, we are friends, then because I can't take a look at a site without starting to draw a line. That line is either supported or erased by the engineer: "Don't go there, it's a mess!"

Geotechnics, water supply, rainwater draining, is that what you mean?

No, actually I think those subjects are secondary, they come about later on. I am talking about the general design of the building, its structure and a coordinator for all those disciplines that make a building sustainable. Today's buildings cannot overspend resources. That is why I cannot draw a distinction now.

Do you think that this relationship has become closer due to these new requirements?

I'll even be more radical than that: I think that in a few years there will be no separation. It is a close, everyday relationship.

I am making a mirror for the Venice Biennale and I am not going to risk making a 20 m mirror that might fall over. Rui Furtado [note: a Portuguese engineer] told me how to do it straight away, what the structure will be like, if it is going to be visible from the sides or not. The relationship is so close that when he is done with the structure, the design will be almost complete. Without the structure, I cannot have my mirror, then either he will sign it or he won't. Things used to be like that. Borromini and Bernini were not architects. They were master builders who knew about everything. The building system ruled.

Do you think that this new relationship, that could be more restricting, is actually more liberating?

I could not build the Braga Stadium without 20 engineers around me and they could not build it if they had not had 20 architects. We do have some intuition, but it is not enough to achieve a good result.

We are all in the same team, one does the cutting, the other is an anaesthetist, etc.

I think I only draw at the final design stage, over the engineering drawings. There is a story I usually tell about the Braga Stadium, a real story. The deadline was so short that there was no time to build and design. The construction works were about to halt, I would make a drawing, a collaborator would build a model, then I would propose some openings in the beams and the walls, some squares or some rectangles. When I saw the engineering models, I would say that those shapes were unacceptable… so the rectangles and the squares would become circles, but how big?

I would then draw elevations with that dimension. I would finish the design and, at night the engineer would design the reinforcement and the design would be at the worksite by 9 AM. This is an example of what I believe the future of construction will be like… Four hand duets!

Is building design moving towards an integrated methodology between the architecture and the engineers?

I understood that many years ago when I went to London for a meeting with Arup. The project I was working on with Siza was the Hanover Pavilion. Over here I cannot only work in that way except with a few engineering offices.

It was an open-space office with cubicles and tables. At the table there was the acoustics specialist, the structural specialist, the safety specialist, etc. When I brought up a topic, each one of them would tell me everything I needed to know about it. The bottom line is that the buildings are well built.

Construction Research

Since nowadays there are so many building regulations to follow and there are more and more building requirements, in what way does research contribute towards greater architectural freedom?

This is where the issue about time comes in: Nowadays, Portugal is a strange country. Designs must be completed by tomorrow and there is no time to think about solutions, unlike the rest of Europe and all over the world. Nowadays, there is not enough time for good architecture and engineering design.

Specialization makes no sense in architecture …

It doesn't make any sense at all. An architect is someone who knows a little bit about everything and he doesn't know everything about anything…

Architecture works this way: you don't have to be an expert in a field. You have to study it, be sensible and then do it. It doesn't matter if you are building a bank or a hospital. You must be in control of the problems.

More and more different disciplines are involved in design and the designers' responsibilities regarding compliance with regulations are growing. Can it be an obstacle to the authors' creativity? Do you think there it will interfere too much with the author's work methodology?

There is a lot of interference but I believe that these limits have never inhibited our imagination, quite the opposite. The more censorship there is (as long as it is reasonable, not just a tantrum) the more well defined architecture becomes. An artist's biggest despair is having total freedom. Architects have a great advantage here. Things are well defined.

I think that among all the information from regulations, some of it is ridiculous, others are intuitive. Regarding sustainability, for instance, some things are completely ridiculous. Architecture is not good because it is sustainable. It is sustainable because it is good. It has to be.

I do not think that there is such a thing as a beautiful building people die from heat in. Sustainability is a matter of common sense. Sustainability is a necessary but not a sufficient condition.

I think that all of that information regarding sustainability is interesting but we cannot be fundamentalist about it.

In Madrid's airport when there are about 5000 people inside, everything is very quiet. It is sustainable. Not a ray of sunshine goes into that airport, but it is bright inside, it is acoustically comfortable. It is about 1 km long, with doors every 200 m, there are about 300 people at each door but there is no noise. People tend to say it is very good because the acoustics are fantastic. I disagree; it is very good because it is an integrated design.

So you believe that the extra technical support architects have today allows further creative freedom. I am thinking about Herzog's work or even Rem Koolhas', although that is more evident in Herzog's work. This is only possible if the architect and the engineer work together.

Many of the building systems that are being adopted were invented by architects. Take the use of nets, for instance. Koolhas applied a net to the Prada Museum then Herzog used it and now they are everywhere. So the magician hides one hand and reveals the other.

I do not see a dichotomy here at all. This is how I work, this is how the good architects I know work. The separation between engineering and architecture does not exist.

Do you think that the depth of the knowledge about building systems, supported by engineering, helps the development of the lucid architectural design you often mention?

Besides that lucidity, you have to find the right solution, which is the difficult part. Architecture is always a game between information and form. The more the available information, the better the resulting form.

Siza draws and he says that drawing is intelligence's desire. Intelligence is a more permanent state.

I just wish for lucidity for that particular moment. This lucidity is a momentary conciliation of the information that is available and the form that is possible.

What does architecture give to construction?

Then there are other details: architecture can't just be the answer to a problem, that is called construction, not architecture. Architecture is construction plus some added value which is creating sensations that make people feel good. It can never be premeditated, if it is, it is a disaster.

So architects must understand how to create objects, as well as possible, and then they can add the details.

Borges sometimes injects some defects in the text, to disconnect it, so that isn't so cohesive and so perfect. It adds some freshness and some life. The same happens in architecture.

When I visited Siza's house in Belgium, it was perfect, rational, anonymous. All of a sudden, you notice a 1.2 m window you could only crawl through which is there so that you could sit down admiring the scenery in winter. That is the unexpected detail.

If you listen to Miles Davis' Jazz, he has a theme that is repeated with some variations throughout the song but, when it seams that he will follow this metric until the end, that the song will finish that way, he does the opposite. Like that corner by Siza… That is the unexpected.

Those gaps and differences can be found throughout history and time.

People like old houses because, although they are less comfortable than newer ones, they have these variations, these unexpected details. Exceptions, no-man's-lands, basements, attics, corridors where people could play football…

I noticed that in your early work, there was careful research about the building system in every project. The study of the building technology that was present in each case was an important tool for the development of the building design. I am talking about houses. In the North, the houses are made in stone, in Alcanena they are made in brick, in the Algarve they are whitewashed…Now that you are at a point in your career where the projects are much larger, is that kind of research still a work tool?

It is, but just as far as it leads to the rationalization of the whole process. I am not saying that this is the way buildings ought to be built, but this is the way they should start. It allows me to find standard measurement and module settings for the materials and to identify the North and South facades, etc. I use this kind of research. Just because I am building a hospital in the Algarve, I am not going to use whitewashed walls painted in blue or use limestone in Lisbon.

That might be done with a smaller object, a *fait divers*, a demonstration of specific research. The important part is whether you are in the right mood and the right dress for the party. So there is another kind of research.

I think I never stopped researching. I started to research different things because the scale of the problems required me to. That is the difference between a pediatrician who treats children and the physician. The scale is different.

Is the Burgo an abstract urban object made with a granitic skin?

It is an old object. It is not something I could do today. It was made when I stopped building just houses. I was in Switzerland at that time so, it must be said, I was influenced by Swiss architecture.

I understood that modernism was exhausted. It couldn't be used in its pure form. Its time was gone. No one believed in pilotis or in a house as a living machine anymore. But that was market language, domino structure. It was the modern language that was not convincing. Post-modernism was even less convincing. It was bizarre. This was quite embarrassing for architects during that period.

This explains the success of Swiss architecture. They did not neglect their cultural traditions. They merged them with modern architecture. They created a kind of hybrid architecture, mixing tradition and modernity. For people from my generation, young people without a well defined language, it was important.

I had no experience in public buildings or large buildings. Burgo was my first large building. It brings up many of the subjects that were mentioned at that time and that were criticized with irony: "the skins and pictorial materials fashion." These are, however, perfectly adequate for a building of that size.

I even explain it at conferences and in articles. The firemen defined the height, the width was determined by the engineers, etc. That was what happened! Then I designed some makeup: some rimmel, some red, blue and pink lipstick…

Then I had meetings with the promoters (BPI [note: a Portuguese bank] at that time). "Why won't you build it in granite? This is a solid bank and we are in Oporto. Or build it in glass because we are a modern and transparent bank. Or in steel because it is high-tech…" I knew that was possible because I couldn't move the floors but I could dress it in a Lacoste polo shirt, a shirt or a T-shirt…

I understood that the skin was a reality. It was not a fashion issue. I took this arbitrary factor to its last consequences.

The building has no base, no body, no ending. You could remove three floors and it would look the same. You could add five floors and it would still look the same, so it is anti-classical.

The Burgo could only be built if it was a lot cheaper than it was designed to be. The structure was there mainly to support the façade, which was built in solid stone and steel. The new clients (who were very nice) said, "We cannot build a structure to support a façade, you will have to make it lighter."

Is it an abstraction? Did this repetition make the building almost anonymous or did it make it sober but not anonymous at all? In the case of the Burgo, you can tell that the abstraction works. There are windows inside.

It does. The façade is sustainable. I mean, the north façade is not sustainable because it is decorative. The window depth in the South façade works very well because they are brise-soleil windows.

The eastern windows are 20 cm high. They only allow direct sunlight to enter at a certain time of the day. The same thing happens with the West facing windows. There is no problem with the Northern façade and the method works perfectly with the South one.

In this building, the design process was followed obsessively, from the first conceptual pictures until the façade detailing… Do you think that the rigor in this process is a quest for the authenticity of the object or the authenticity of the concept?

Burgo is an authentic building because it is a mirror of the lie it really is. It was not by chance that the door appears to be a rabattement of the building planes and it shows that everything is a phoney. That is why the word "Burgo" is written the other way round and why you can see the concrete columns there. It is not a stack or it would not have columns.

It is like the theme song from Truffaut's movie, "La Nuit Américaine". At the end of Truffaut and Mel Brooks' movies, the cameras are moved so that the set can be seen from behind. You can see people drinking a Coca-Cola or Julius Cesar smoking a cigarette, dismantling the plot and showing that "This is fiction"

I felt the need to admit it: "Lookout, this is all a lie!" so at the time I decided to confess and to tell the truth… "It is a lie".

Is that door, with the letters facing the wrong way, an obvious escape or is it a composition option? Those letters are only facing the wrong way because that wall is a rabattement of the façade…

It is a finishing touch. It enhances the idea that it has been pulled out. It should be part of the wall and it is made out of panels. By pulling out that panel, people have a perception that there has been a rotation of that section. People entering the building can see the columns and the beams. Actually, it is not just formalism, for two main reasons. Up to the fourth or fifth floor, the columns are so slender that horizontal steel elements were necessary. Many of those brise-soleils are real. They are actual structural elements.

Those horizontal boxes exist up to a certain level, then I kept using the same elements, but they are false.

I needed brise-soleils anyway. I would either use window shades, which is not easy when the windows are 70 m high, or I would use static elements that would create that overlapping effect.

Regarding the relationship between being anonymous and being sober that you often mention, what do you think about the urban impact of this building in Avenida da Boavista?

The word "anonymous" is very dangerous because it can be perceived as being very snobbish… The Burgo is not anonymous. I often say that the ultimate goal for an architect is to become anonymous. After is ready, the building is no longer ours. This means that when there is a collective adherence, the building ceases to be ours and it becomes everyone's. This is the ultimate goal for an architect.

When people no longer say that a building is by someone and start to say that the Belém Tower belongs to Lisbon or that the Clérigos Tower belongs to Oporto, the buildings become anonymous. This is my goal and it is every architect's wish: for a building to become everyone's, not ours.

Construction Design

Was there a big change in the project when the construction work started, 10 years later? Did the designs go back to step one?

The design did not change too much. The most important issues were the costs and choosing the materials … The showroom itself became a 1:1 scale model. I had a new goal. I had to achieve that cost or it wouldn't be built. That is why different materials and detailed designs were needed.

In the first design the granite walls in the façade were 8 cm thick…and now they are 2–3 cm thick. I evidenced this fragile look with bolts. I didn't do any pasting, I did not want it to look like a stone wall at all. It was not thick enough.

The structure was also redone. The previous design had thin concrete slabs but they were made lighter, molded slabs. Of course, this changed the original layout completely. Everything had to be redesigned and reconfigured because the modules are all different. The elements are not the same.

All of the floors have different heights, the elements are different and so are the gaps between them.

It is something you also learn.

Were they the result of the stack shape that was adopted? How did these functional aspects interact with the architectural design? How do they relate to the architectural design?

It has a solid structural core that frees the building from columns. It is an open-space because I did not know what would happen to it. In an open-space the columns must be moved to the façade, which presents an important advantage: the structure determines the façade's design.

I couldn't face the building westwards towards Avenida da Boavista. It would be unprotected. In the late afternoon the sun would shine on the Western façade and nobody would be able to see at all. I had to move the rooms to the North and to the South. They were fine to the North and I had the brise-soleil to the South. All the bathrooms, kitchenettes and corridors were placed to the East and to the West of the building. Since there are windows, the corridors are properly lit.

This was when the façade system with 1.10 m opaque strips and 0.20 m high windows appeared on the Eastern and Western façades. The South and North façades are the opposite: 1.10 m high windows and 0.20 high strips.

The solution is perfect because it is an architectonic language. The design was adapted to an important requirement.

In these façades, we can see all the principles of shading. There are walls to the West and window overhangs to the South…In a bad design, recommendations do not blend in. Good design integrates sustainability in the work itself, improving it.

This brings us to the beginning of our conversation again… Although there are work teams that support the architect's work, he must have a deep understanding of the technological aspects so that he can reflect them in his design since its earliest stages.

Architecture must design and create forms, and in order to do so it needs information. This is one of the few rules in architecture: information/form. Architecture is not autonomous. We need a lot of information to create adequate forms. This is called construction.

Architecture is not just this information. It is the information plus some added value. Up to a certain point, there are rules that are developed using information. From there on, you have to know how to deal with them.

General Arrangement Drawings

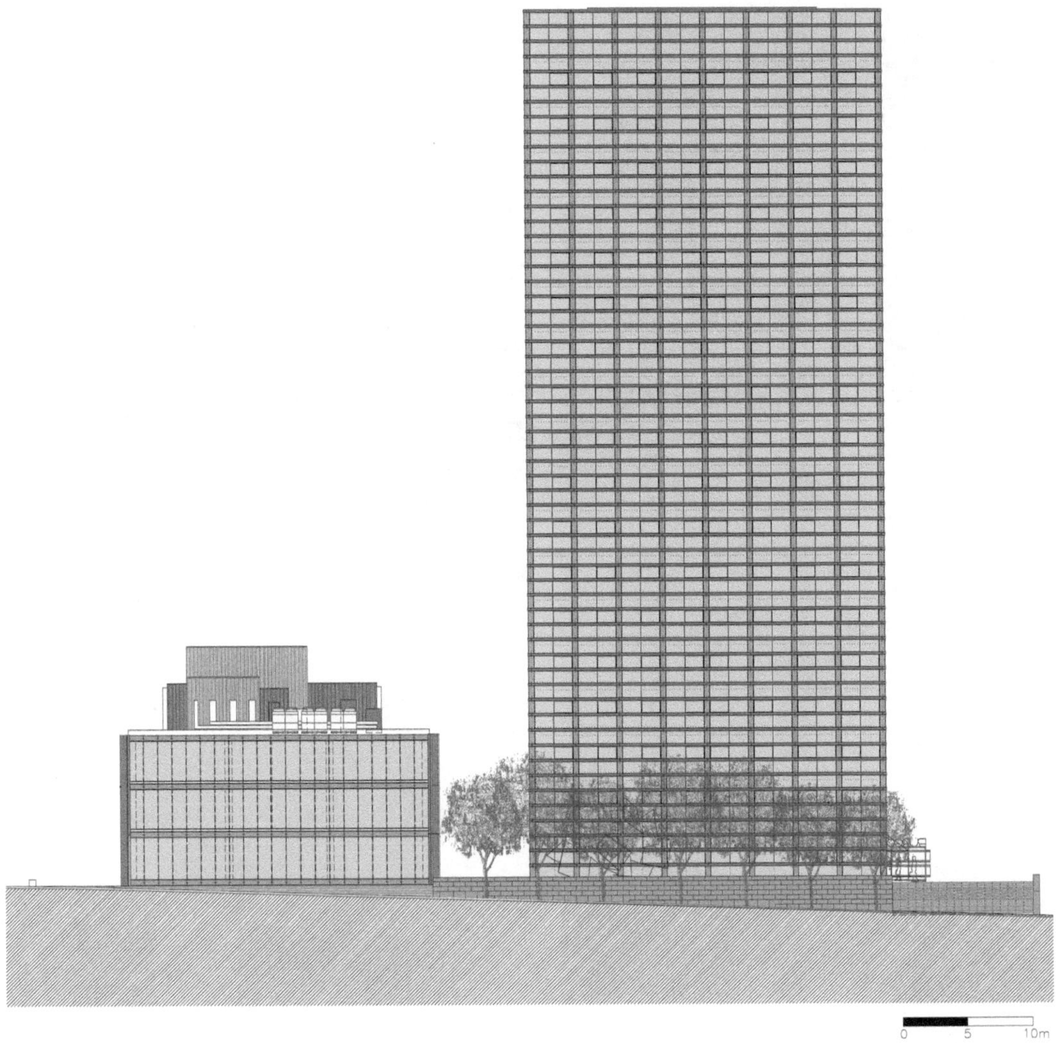

North façade

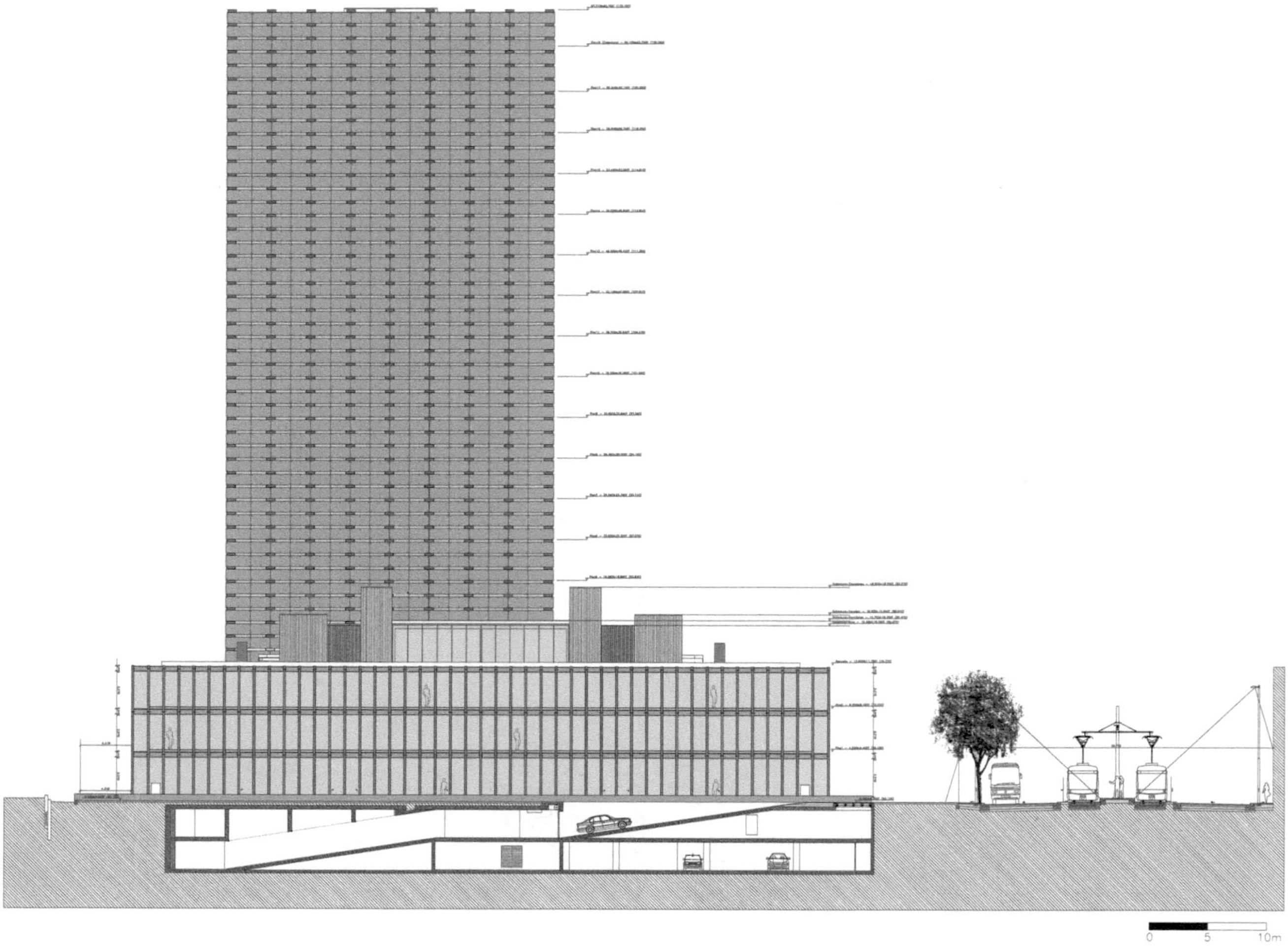

East façade

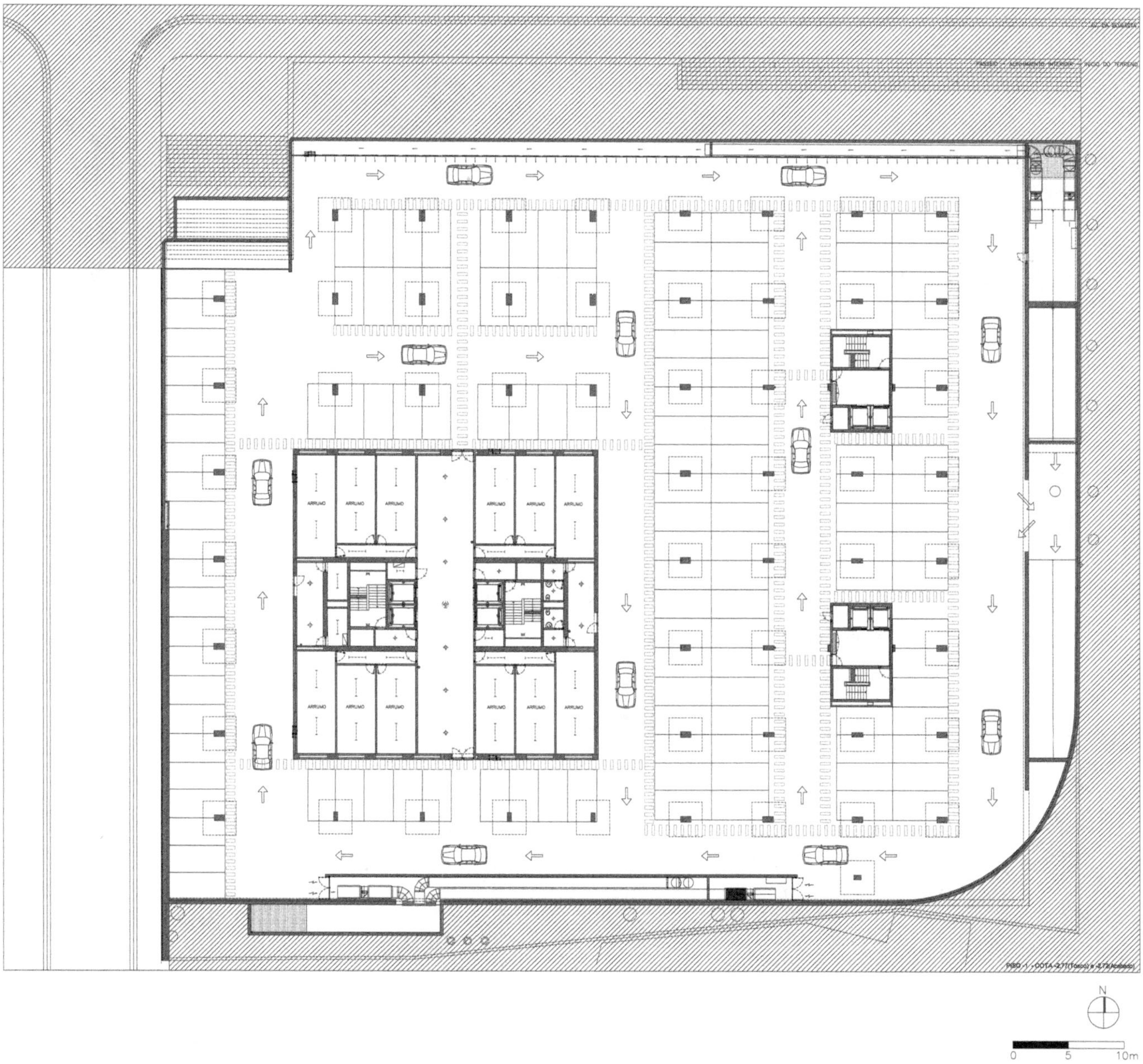

Plan of Level-1

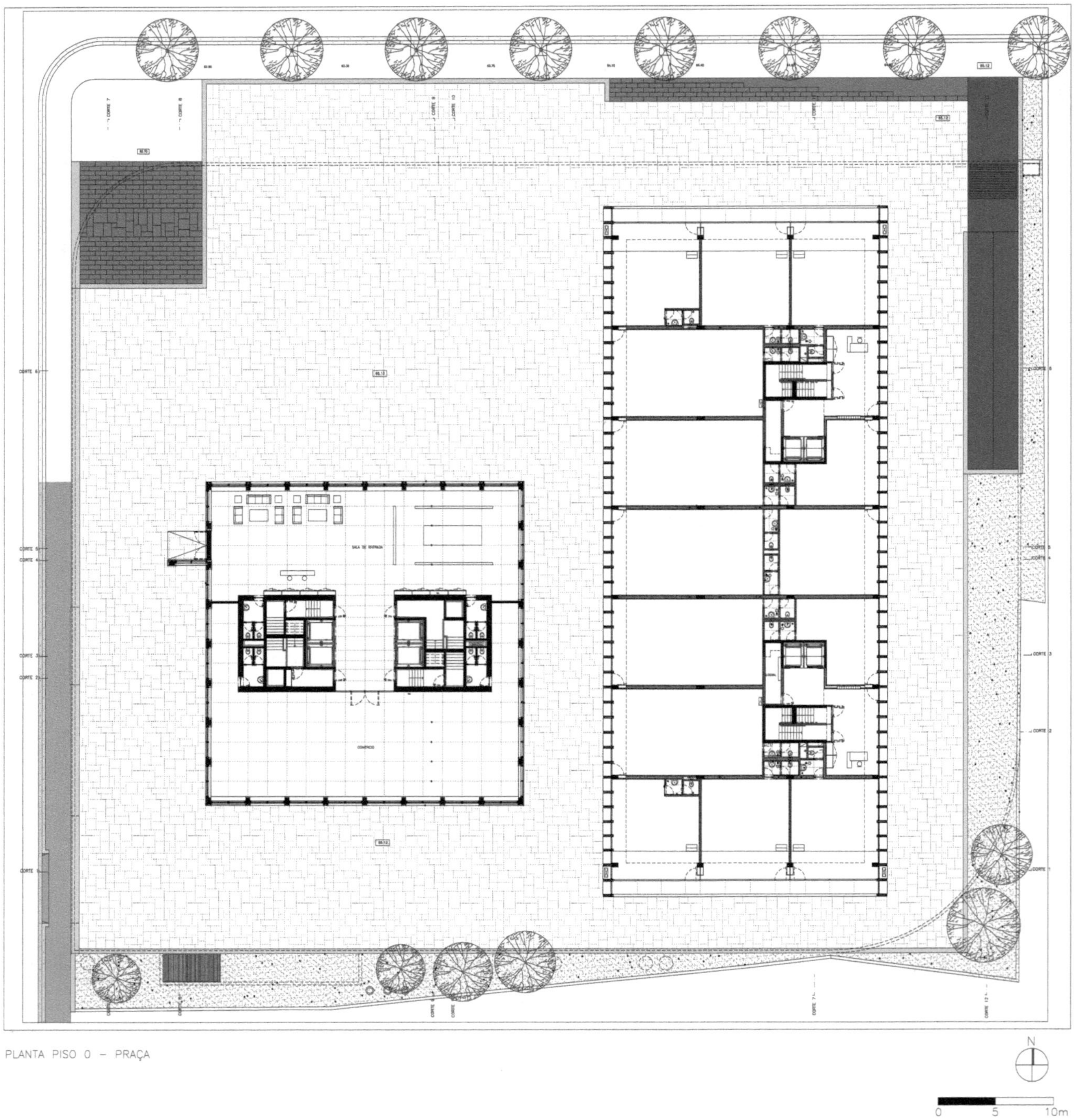

Plan of Level 0

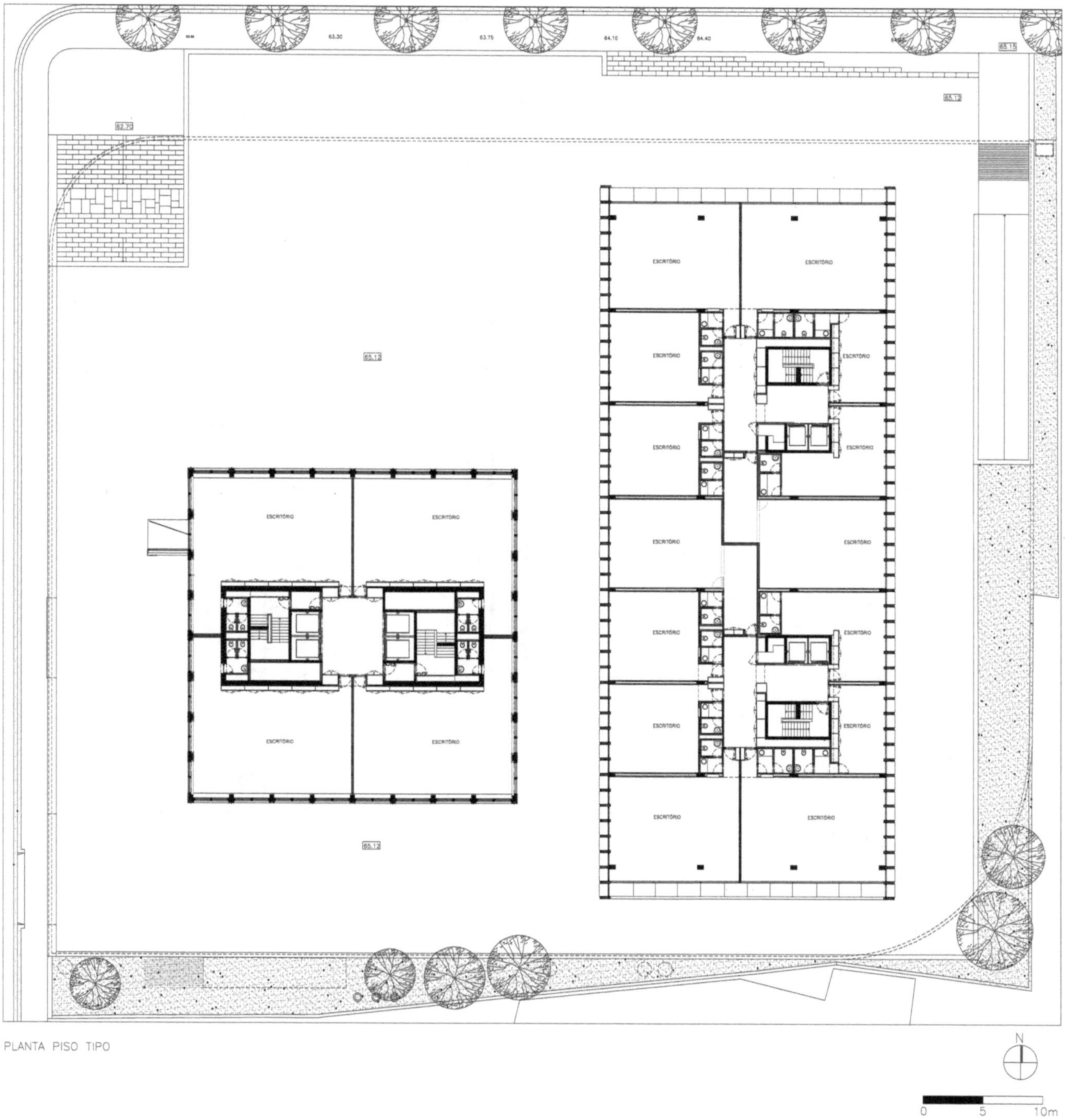

Plan from Level 1 to Level 17

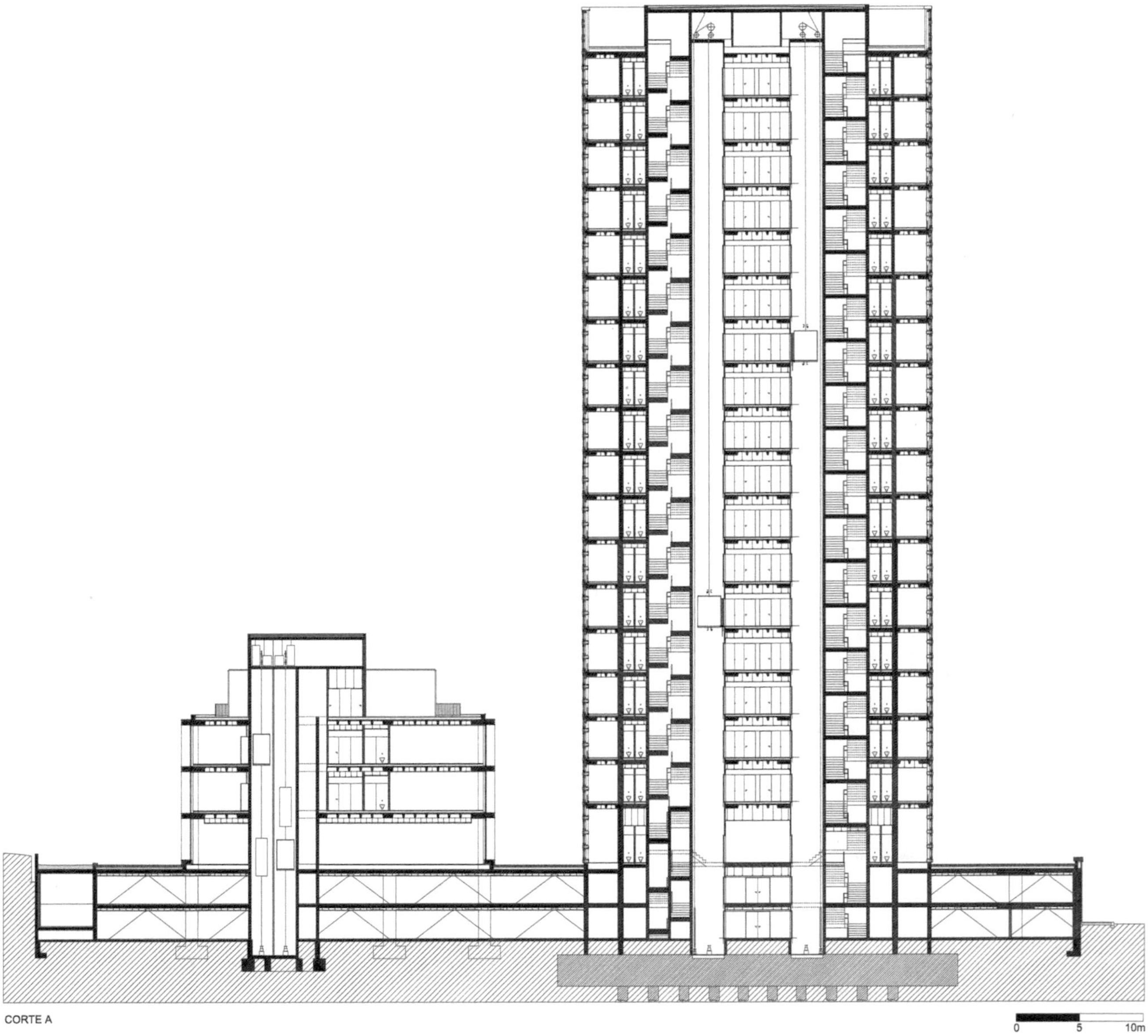

Section A

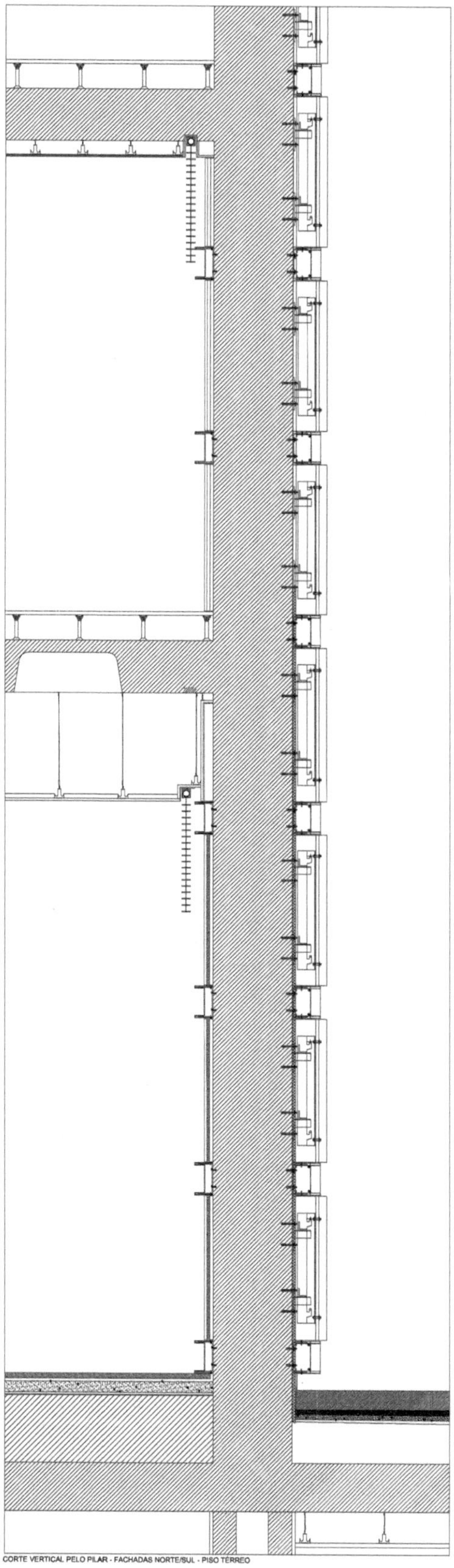

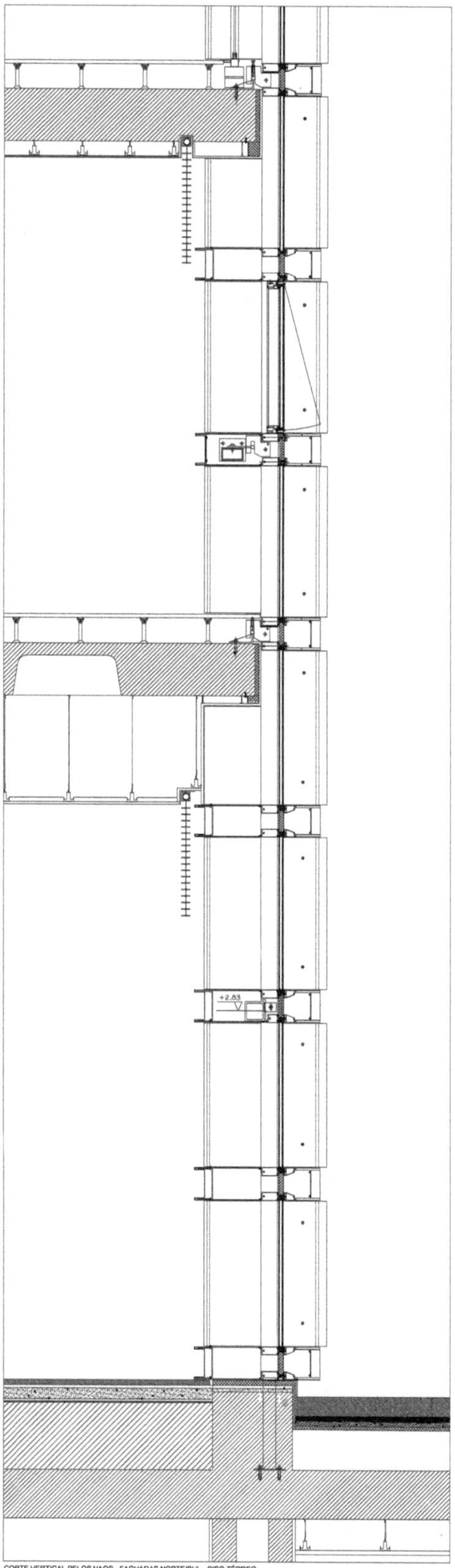

Vertical Section of the south and west façade
Vertical Section of the north and east façade

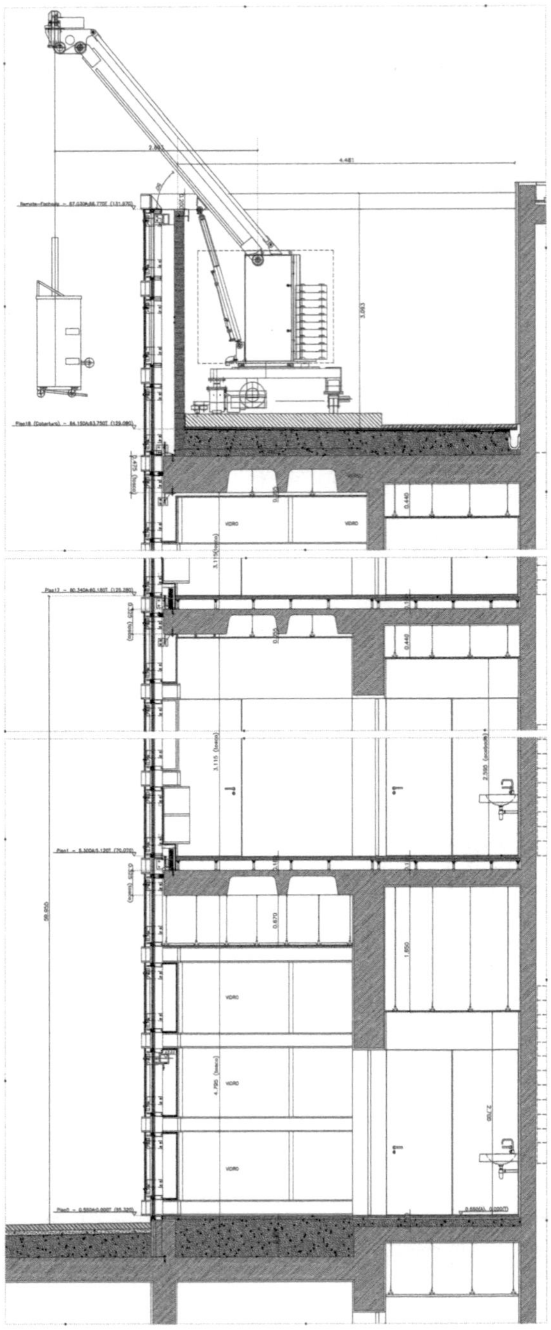

Vertical Section of the south and west façade
Vertical Section of the north and east façade

General view Copyright Luis Ferreira Alves

View of the corner of the two types od façades Copyright Luis Ferreira Alves

Image of the interior facing north Copyright Luis Ferreira Alves

Image of the interior facing west Copyright Luis Ferreira Alves

Building Facades Prefabrication. The Lesson of the Burgo Building

José Manuel Amorim Faria

Object

We explore the idea that prefabrication is not a prison but rather a source of inspiration in architectonical creation. The Burgo Building is used as an example to explain that idea.

My primary purpose is to demonstrate that prefabrication implies simplification that allows architects to *explode* simple ideas, organizing and disciplining them, in order to accomplish a functional objective goal.

Further, I contend that increasing the use of prefabrication implies the adoption of a modular and structured reasoning and the respect of constructive tolerances and constructive joints of the various construction elements and systems.

Prefabrication and Architecture

Since the most early times, people have been searching for means to perform basic tasks in a more or less repetitive way in order to reduce the amount of effort and energy needed to do them.

It is possible to find examples in many areas, from Agriculture to the confection of meals.

The pursuit of increases in productivity (reduction of the amount of time spent doing a unit of a certain task), or, expressing it in another way, reducing the effort spent to achieve the same goals, has always represented a major concern of mankind.

Therefore, at the Construction level, industrialization has been assumed as a fundamental solution to the problem of how to accomplish this essential basic goal.

Prefabrication represents one of the roads of construction industrialization. The other main two ways to achieve it are rationalization[1] and mechanization.[2]

Currently, prefabrication is defined as "a set of construction techniques that are based on the production of construction elements outside of their final places of definitive setting, on site or in an external production unit, which are afterwards connected and assembled on site".

Prefabrication implies a certain discipline on the conception related with geometrical organization and repetition which also has always represented a barrier to its penetration on the fields of higher importance to Architects.

Therefore, in general, architects have always considered prefabrication as an enemy to liberty of creation, a *prison* to which they refused to submit. For an Architect, however, prefabrication may constitute a motivation and at the same time a paradigmatic inspiration of his or her creative process, as the Burgo Building case very well exemplifies. So, the prison becomes liberation and the geometrical order and simplicity contribute to the explosion of any original simple idea. The Building becomes object and form but also function and service, assuming clearly a representativity that is an outcome of its apparent esthetical simplicity.

The Problems of Prefabrication [1]

The creation of prefabricated systems with a high level of prefabrication implies the adoption of modular dimensional coordination, which means submission of architectural volumes to a normalized dimensions system, following the three

[1]Construction Rationalization – set of organization actions, planning and verifications that contribute to an increase of the productivity of the construction sector as a hole or of each of its individual areas and tasks in particular.

[2]Construction Mechanization – Substitution of manual labor for a machine as a way of increasing productivity and quality of construction.

J.M.A. Faria (✉)
Faculty of Engineer, Oporto University, Porto, Portugal
e-mail: jmfaria@fe.up.pt

V. Abrantes et al. (eds.), *The Pre-Fabrication of Building Facades*, Building Research: Design, Construction and Technologies, DOI 10.1007/978-3-319-22695-8_2

Cartesian referential axes. Therefore, Modular dimensional coordination involves a definition of modular dimensions (multiple or sub-multiples of the standard module) of the different parts that compose a building so that its assembling and connection may be performed during construction using the minimum possible of resources.

The basic module adopted by ISO (International Standards Organization) is 10 cm and the most common multi-modules are 20, 40 and 60 cm.

In general, the systems of modular dimensions adopt values that refer to the central axes of the parts and, within the limits of the building, refer those same values to its external faces. All the dimensions must be considered in the final solution, including all finishing.

It is not possible to adopt prefabrication in an economical and efficient way, mainly on facades, if the architectural conception is not fit to a modular system of dimensions.

This constitutes the first and more important difficulty related to prefabrication which can be easily solved using, since the beginning of the design process, a modular geometrical referential of dimensions.

The other two main difficulties to overcome are associated with the concepts of construction tolerances and construction joints.

The idea of tolerance is very well known in all fields of activity in which there is a need to assemble parts whose final working success is intimately related to the respect and control of the inevitable deviations associated with the production of its elements. In Construction, the concept applies to the dimension of the spaces and of the components and elements of construction, but also to the setting errors during the construction process on site.

The problem of fixing admissible tolerances to a certain prefabricated construction system consists therefore of solving the following specific problems:

- Maximum and minimum dimensions of components (width, length and thickness) considering the special characteristics of the corresponding production and assembling operations;
- Definition of the maximum admissible variations on the surface planity and verticality of the final finishes.

The tolerances apply to the prefabricated components and represent the basis on which the dimensions and type of the construction joints can be based.

The construction joints are usually divided in three fundamental types:

- Large Joints (1–4 cm) which are able to absorb efficiently all the dimensional deviations related with a component – this is the case of all the structural joints or of the joint between a plasterboard partition and the structural slab;
- Medium Joints of second level (1–10 mm) which are able to absorb assembling and production tolerances of more rough materials and/or absorb variational deviations associated with thermo hygrometrical origins, such as wooden floor coverings, glued ceramic tiling of external walls or joints between masonry elements;
- Small joints of third level (0–3 mm), clearly exclusively used to absorb small production tolerances of components with a high level of quality control on the production operations.

The construction joints have to serve many different and important functions such as air permeability, water permeability and mechanical resistance.

It must never be forgotten that each construction element, after assembling and erection, includes all the different components and the corresponding joints and that the performance behavior of the element is related with the global functioning of components and joints set on site and not only to the individual performance of each one of the two parts.

The three above mentioned aspects represent the basic fundamental principles to be respected during the design process of prefabricated construction elements and must NEVER be disregarded.

The Burgo Building case [2]

The Burgo Building represents an obvious illustration that it is possible to design a building with a high level of prefabrication and, at the same time, give it a very strong character and personality.

The initial creation of facade systems is inspired by simple human objects and tools (Fig. 1).

In a second phase of the process, the modular geometrical references are used. The building as a hole respects a modular design (Fig. 2).

Afterwards, we can acknowledge a design of joints that is adequate to the case of the designed facades.

External solar protection construction systems are adopted.

A detailed hierarchy of structural supporting elements is defined (reinforced concrete piles, horizontal steel beams, aluminum and glass external structure, solar protection system supporting elements).

The volumes are fit to place.

The best setting on the site is studied, considering solar exposure. Different kinds of facades are defined according to their geographical orientation (west well protected, east and south a bit less, north without protection).

The construction systems are studied and all the joints are well detailed.

Fig. 1 Primary sources of inspiration of the Burgo Building

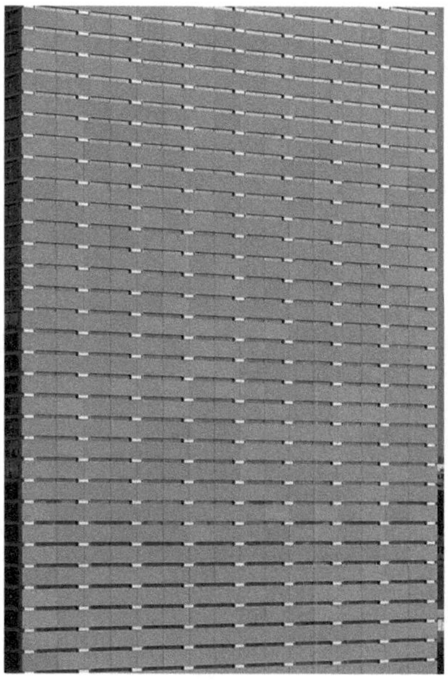

Fig. 2 The modularity of the Burgo Building viewed on its facades

The design introduces bridges between the internal spaces and the envelope. The architectural design is liberated when it goes to the internal spaces because the prefabricated envelope has been totally conceived and has no major difficulties of communication to the inside. The internal spaces can now be studied freely, respecting the program.

Prefabrication helped to give freedom and wings to a very simple original idea.

Conclusion

This paper has been written by an engineer with a Ph.D. on lightweight prefabrication. It represents a somehow daring appearance in a cloudy area, placed somewhere between Architecture and Civil Engineering, that Architects in general in Portugal call *Construction.*

It aims to induce the architects' appetite for prefabrication, illustrating with metaphors, photos and images its main advantages, without forgetting to present in parallel the main difficulties to be solved. It is hoped that it may have contributed to a more realistic and liberating idea of prefabrication, analyzed both as a tool and as a methodology of the creative process.

References

1. Amorim Faria, José (1996) Divisórias leves prefabricadas – Concepção e análise da viabilidade de um sistema realizado com base em madeira e derivados – Tese de doutoramento – FEUP, Available for download on www.fe.up.pt – library – cathalogue – only in portuguese. http://paginas.fe.up.pt/~jmfaria/Publicacoes1_75/Dissertacoestesesesimilares/9/Divisorias%20Leves_9.pdf
2. CdO – Cadernos d'Obra – n° 1 – January 2009 – FEUP Edições, Porto. https://paginas.fe.up.pt/~cdo/index.php?/projects/01/

Design of Façades

Hipólito de Sousa and Rui Sousa

Introduction

In spite of the importance that the vertical envelope of buildings has always had, nowadays unquestionably a greater value is given to image and to an attempt to make the most of the potentialities and bold use of technique as well as materials. In recent years this greater emphasis on image and on a set of requirements concerning comfort and durability in particular, has witnessed the need for specialised technical involvement in new areas such as the envelope of buildings.

New expressions have appeared in the technical world such as "Engineering of Façades", in an effort to try to make those involved in this area more aware of the importance and difficulty of this element in the construction process. In fact in many buildings, finding the most suitable concept for façade solutions leads to complicated technical problems requiring multi-disciplinary and highly specialized contributions.

The aim of this article, which uses as its main reference the "Burgo" building, a background image in the whole magazine, is to highlight the importance of window frames. We pay careful attention to technical methods for developments in this area, where a perfect union between Architecture and Engineering is essential, as well as thorough knowledge of engineering based sciences, materials, physics of constructions and technologies.

Brief Description of the Envelope of the "O Burgo" Building

Generally speaking, the envelope of the building in question has two types of solution: the first in its north and south orientations, and the other in the east and west façades. The first is composed of a system of curtain walls, self-supporting and made up of aluminium cross- beams and struts, which are totally or partially visible from the outside, and modular panels of glass and natural stone, fixed to the structure of the building (Fig. 1). In this façade, as well as the stone panels there are moving windows (projecting) and fixed parts.

These modules have approximate unitary dimensions of 3×3 m^2, limited horizontally by its slabs and vertically by its columns.

In the stone façades facing east and west, the main panels are made up of granite cladding, with rear insulation and interior finishing, fixed mechanically to the self-supporting structure of the façade. The cross-pieces of the curtain wall are partially interrupted by small dimensional glass panels (about 20 cm in height) with aluminium frames.

The method of construction of the façade consists generically of a main aluminium structure made of large dimensional rectangular tubular beams, and of a secondary structure with beams that support the assembly of the diverse systems and construction components of the façade (ironwork, glass/stone panels, water-tight seals, etc.). The curtain walls are fixed to the concrete structure of the building (columns and floor slabs) by means of mechanical fastenings (Fig. 2).

Behaviour of Glass Façades

Main Performance Requirements

The façades of the buildings are subject to a set of performance requirements that influence the behaviour of the buildings in various aspects, ranging from acoustic comfort, thermal comfort to energy efficiency, stability, safety in use, environmental hygiene and health, among others. Compliance with these requirements guarantees an adequate level of comfort and safety, and reduces the incidence of pathologies and also provides energy economy [1].

H. de Sousa (✉) • R. Sousa
Faculty of Engineer, Oporto University, Porto, Portugal
e-mail: hipolito@fe.up.pt; ruysousa@fe.up.pt

V. Abrantes et al. (eds.), *The Pre-Fabrication of Building Facades*, Building Research: Design, Construction and Technologies,
DOI 10.1007/978-3-319-22695-8_3

Fig. 1 General view of the building and partial view of the curtain wall

The glass components of the façades are subject to recent standards [2, 3] involving a set of requirements that are mandatory for these products to have in the scope of CE marking. In the case of doors, windows and curtain wall, it is necessary to characterize the water-tightness, air resistance, wind resistance and impact loads, thermal insulation, and noise isolation, among others.

This characterisation is indispensable for an appropriate evaluation, selection and use of glass façades in buildings.

There is an informative document produced by the National Civil Engineering Laboratory, ITE 51 [6], that defines minimum expected performance levels of doors, windows and curtain walls in the buildings, in various conditions of exposure to the wind and rain in Portugal.

This document is based on Portuguese regulations (RSA, RCCTE and RSIEH), European Standards that are applicable to the frames and their components, as well as benchmark French recommendations for window frames. This document provides assistance with regard to the mechanical design and choice of windows, doors and curtain walling systems, in order to assure a satisfactory minimum performance in aspects related with watertightness, air permeability, wind resistance and resistance to mechanical actions related to the use of these systems.

In this way, for a specific situation of exposure of the building facades, it is possible to evaluate the aptitude to usability and the level of quality of these systems by means of a comparison between the performance classes obtained from laboratory tests and the minimum expected classes obtained from ITE 51 [6].

Characterization of the Performance

The performance of the façades with regard to watertightness, air and wind resistance is characterised by means of performance levels. These levels of performance correspond to the boundary values of the pressure/drop in pressure associated with the action of the wind in the buildings, and can only be determined through laboratory tests performed on real size samples.

In the case of watertightness it is possible to carry out a “in situ” test. This test is normative [4] and is carried out without the influence of pressure, by submitting a given area

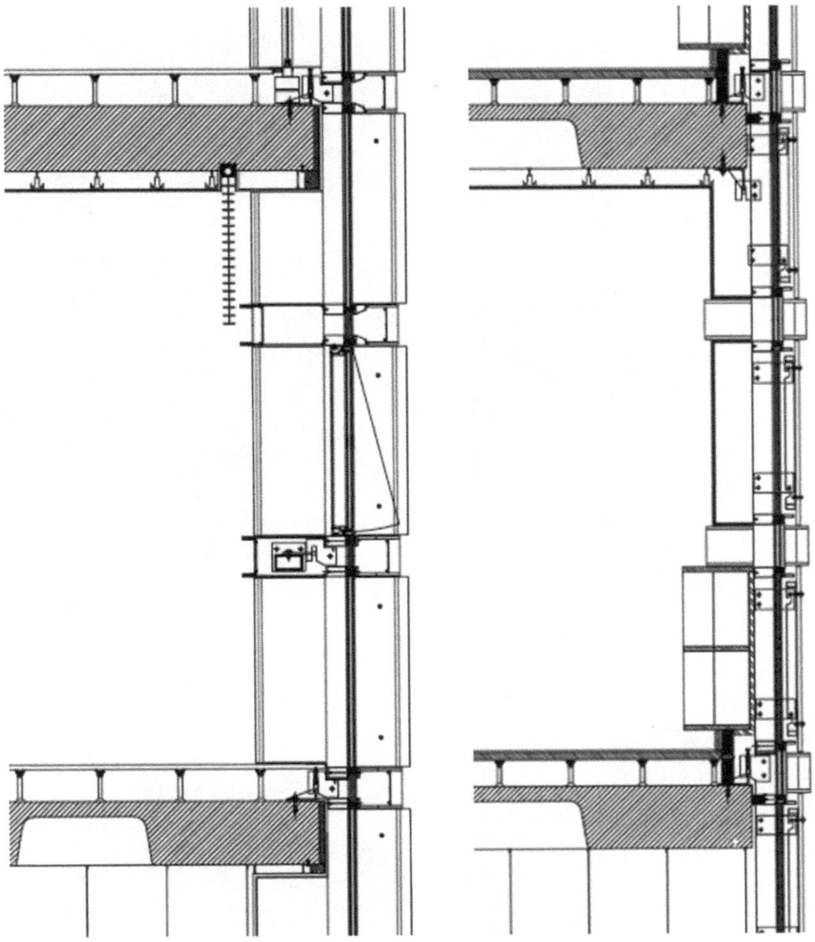

Fig. 2 Vertical cut in the zone of the glass façade (*to the left*) and in the zone of stone façade (*to the right*)

of the façades (fixed and/or mobile zones) to the action of a continuous film of water. This is a simplified test to complement the test carried out in the laboratory, and aims to determine the loss of watertightness associated to on-site construction errors of new or used façades, and its objective is not to classify performance.

FEUP, namely the Laboratory of Systems and Components (LSC), has carried out several studies in this area, aiming at determining the main characteristics of performance of doors, windows and curtain walls systems, in particular the characteristics associated to watertightness, air permeability and wind resistance, as well as support in the study and development of new systems (Fig. 3).

In a general way the experimental characterisation of these systems in the laboratory conditions is carried out in the following way [5]:

- assembly of the test sample (window, door or facade module) in the testing chamber by the Installer/Manufacturer, in the same on site construction conditions (Fig. 4);
- carrying out the air permeability test, by measuring the volume of air that flows out of the sample for a specific pressure value, thus allowing to determine the respective performance class in accordance with the benchmark standard (Fig. 5);
- carrying out the watertightness test, where the performance class is assessed through the loss of watertightness observed during the experiment (signs of water leakage in places that should have remained dry), for a determined pressure level (Fig. 6);
- wind resistance test, where the classification is based on a global evaluation of the results obtained from three tests (Fig. 7) [2, 3]:

Fig. 3 General aspect of the frame testing equipment at LSC (FEUP)

Fig. 4 Assembly of the frame in the testing chamber

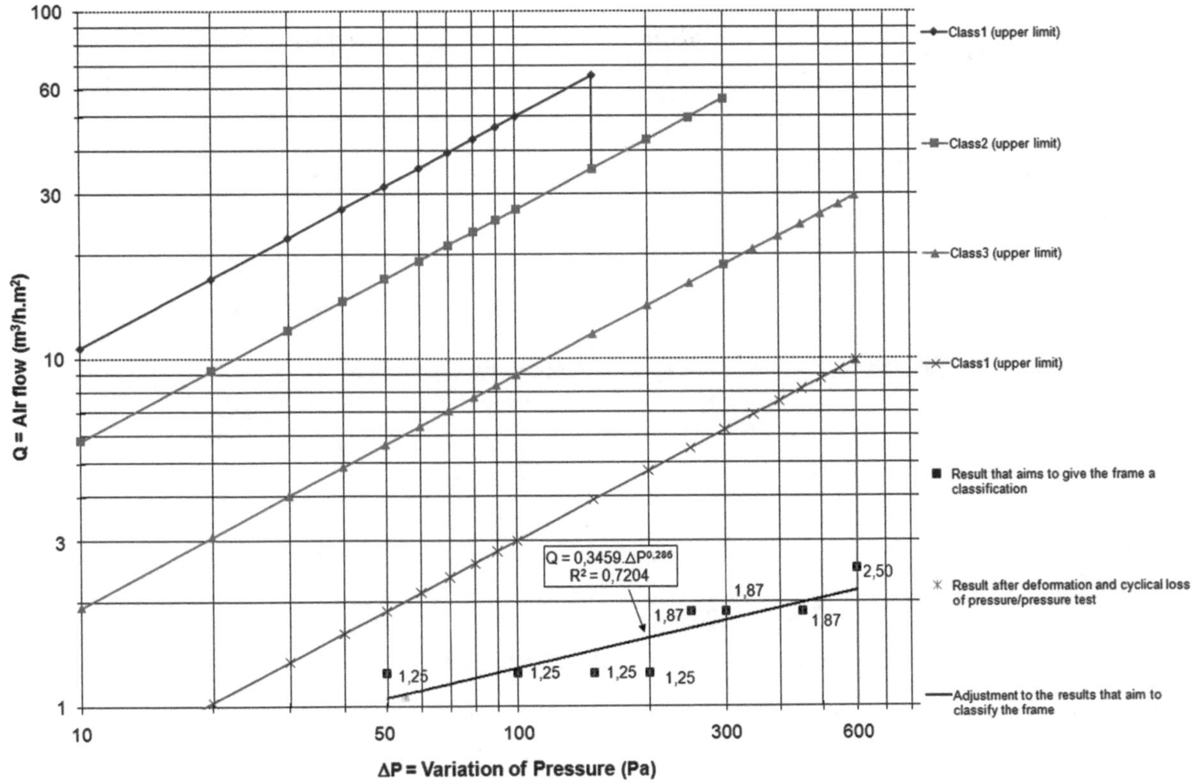

Fig. 5 Registered values and classification of the frame in the air permeability test

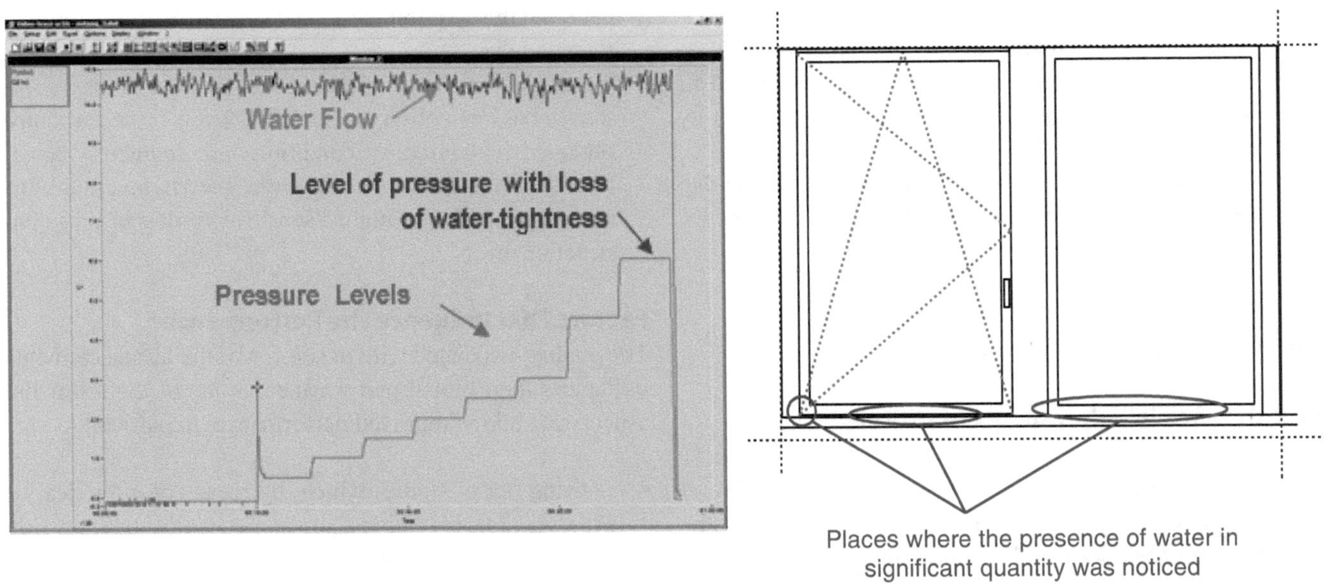

Fig. 6 Example of results of the water-tightness test, with registered test values a table indicating places where loss of water-tightness was observed

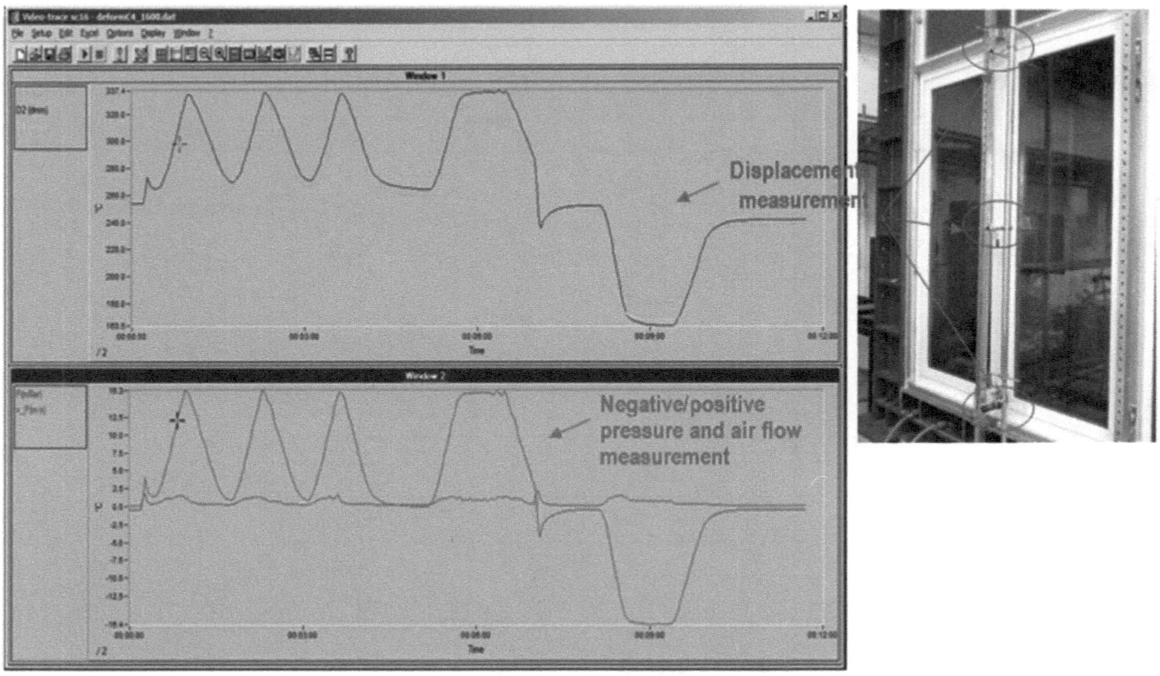

Fig. 7 Example of the wind resistance test – deformation test (register of positive/negative pressure and frame deflection)

- deformation test, which aims to determine the absolute and relative deflection of the components by wind actions in serviceability conditions, thus allowing a comparison with the benchmark values;
- cyclical test, which aims to determine losses of functionality of the mobile and fixed elements and, to determine an eventual increment in the value of the air flow by means of carrying out another air permeability test;
- safety test, which aims to determine the stability and to detect eventual permanent damage on the components, in situations of extreme wind actions.

Analysis of Frame Performance

Performance in Laboratory Conditions

Based on the minimum performance levels found in ITE 51 [6] for window frames, and the results obtained from several tests carried out in the LSC, the following aspects on the performance of windows can be highlighted [5]:

- resistance and mechanical stability are in general satisfactory in situations of extreme wind, with no evidences of failure or permanent deformation of components;
- the deflection of the frame components is frequently imbalanced in serviceability conditions, i.e., it varies between very deformable and very rigid (outside the range of reference limits of 1/150 and 1/300);
- in general, although different performance levels are obtained for negative and positive pressures, the levels of air permeability obtained are higher than the minimum levels required, which seem to indicate, besides the verification of this requirements, an decrease of air loss through the windows;
- watertightness is one of the aspects where a worse performance has been observed since the levels of performance obtained in laboratory conditions are frequently lower than the levels required; moreover, watertightness performance seems to be highly sensitive to design and construction errors.

Factors That Influence the Performance

The quality associated with the design of the facade elements influences their global performance. Some of the following aspects can allow improved performance, in particular:

- carrying out of studies which, by means of numerical or experimental simulations, aim at mechanical optimisation of the frames behaviour;
- choice of frames with a geometrical shape adequate for wind actions in serviceability and extreme conditions;
- definition of adequate channels for water draining and detailed definition of watertight joints;
- sufficient stiffness for the joints and metalwork of the frames, and an adequate connection of these elements to the building (arrangement and number of fastenings);

- sufficient detail given to construction solutions when preparing and assembling the frames, either in factory or on site conditions.

On the other hand, the quality of production and on-site assembly has a determining influence on the watertightness and air permeability performances. Some of the following aspects can allow improved performance:

- quality of the cutting and machine finishing of the metal beams in the factory;
- careful assembly of the components, including sealing of the beams joints, metalwork and the watertight joints;
- careful installation of the infill panels (usually glass panels);
- square edging and positioning of the frames on the building.

Recommendations for Improvement in the Quality of Construction of External Façade Elements

For an improved performance of façade elements, such as windows, doors or curtain walling systems, the following aspects seem fundamental:

- in the planning stage of the project, to specify the performance levels intended for façade elements that are suitable for the intended exposure, opting for the choice of certified construction systems (CE marking or others), or to request the manufacturer for performance indicators;
- to raise awareness amongst those involved in the construction process (designers, manufacturers, installers) about the importance of design, detailing and assembly quality in the performance of façade systems;
- in the production process, to promote more in-depth design and development studies based on specialized theoretical and experimental studies, to supply certificates of conformity of the product (CE marking or others), to produce construction details sufficiently clear and detailed to be used by designers and builders, to use modern production processes with quality control;
- in the design of façade elements or in the design of important constructions/buildings, to carry out laboratory experiments to classify the behaviour of the façade elements in order to facilitate the certification process;
- to assemble of the façade elements carefully during on-site installation, inspecting the critical points and using specialised and experienced workmanship.

References

1. Giacomini E, Sousa R (2008) O Vidro nas Fachadas. Revista Construção Magazine, ISSN 1645-1767, nº26, pp 15–18
2. EN 13830 (2003) Curtain walling – product standard. CEN, Brussels
3. EN 14351-1 (2006) Windows and doors – product standard, performance, characteristics – Part 1: Windows and external pedestrian doorsets without resistance to fire and/or smoke leakage characteristics. CEN, Brussels
4. EN 13051 (2001) Curtain Walling – watertightness – site test. CEN, Brussels
5. Castro M, Sousa H, Sousa R (2007) Caracterização Experimental de Caixilhos em Laboratório. Congresso Construção, Coimbra
6. Carlos Viegas J (2006) Componentes de edifícios: Selecção de caixilharia e seu dimensionamento mecânico. ITE 51, LNEC

The Importance of Design in the Prevention of Accidents in Construction

Alfredo Soeiro

Safety is to apply common sense and the accidents happen when the common sense is absent.

Motivation to Prevent Accidents in Pre-construction Phase

The anguish, the affliction and the sadness suffered by a wife, mother, family and friends in losing a husband or a son or a friend in an accident in construction are traumatic and touch feelings strongly. An image of a person who died or was seriously wounded while working provokes interrogations and perplexities for those who, in a modern world, have no answer or comforting comment from the moral or social point of view.

The world of construction is dangerous and has many potential risks of life and of corporal damages. The exposure to risks of accidents is high and practically constant. For the betterment of society a Culture of Safety has to be created in all phases and processes of construction. The minimum requirements for prevention of accidents should be well known by everyone involved in the construction process. To assure safety in a construction site is a complex challenge. For that reason it is imperative that we pursue a culture of safety. Human life and corporal integrity are precious goods and health is an essential condition of the quality of life.

The activities of construction are, essentially, different from other industries in what concerns the mobility of the workplace and the repetition of conditions in tasks implemented. For instance, the environment of operations in construction sites changes constantly due to weather changes and to location of the site. This diversity is considered one of the main conditions that induce insecure behaviour and that can hinder prevention measures. Therefore it is difficult to legislate and to standardize practices for the enormous variety of size and complexity of construction projects. The variety of construction organizations and of working structures also contributes an additional difficulty for acting, via legislation or practice, effectively in preventing accidents in construction [1].

Culture of Safety in Construction Environment

The current legislation has a character that is essentially descriptive. The development of the role of safety's coordination in the design phase and in the execution phase was created by national laws in each country of the European Union as a consequence of the European directive no. 97 of 1992. The legislation's purpose is to improve the safety and the patterns of health in construction sites. These rules impose a concept of safety and health, based on a linkage of responsibilities that includes owners. These responsibilities in conception, administration and verification of prevention measures devolve onto all who are involved in construction. Owners need to start the prevention procedures in all phases, including the design phase. Safety's coordinators are the nuclei of the articulation of these measures even in the design phase.

Need for Implementation of Safety in Design Phase

Historically, the duties of implementing safety were assigned to contractors as an executioner of the construction work. This is a current situation in other industries. The legislation changed this situation and execution of prevention measures no longer depends only on the contractor but on the workers, owners and designers. This partition of duties is consecrated in the legislation and all participators share part of the responsibility [2].

A. Soeiro (✉)
Faculty of Engineer, Oporto University, Porto, Portugal
e-mail: avsoeiro@fe.up.pt

V. Abrantes et al. (eds.), *The Pre-Fabrication of Building Facades*, Building Research: Design, Construction and Technologies, DOI 10.1007/978-3-319-22695-8_4

Construction owners and designers must assure the safe implementation of the construction works. Legal competence of the safety coordinator is required in the design phase of approving architectural options and choosing techniques to minimize the risks of accidents in the execution phase. It is a difficult task since it collides, in many cases, with the creative and technical options of the designers and, above all, for suggesting more expensive options to the construction owner. But safety and accident prevention have to begin in the design phase as a way to minimize risks for contractors and to adapt the work to those who execute it [3, 4].

Big improvements in preventing accidents while designing can be reached to avoid future risks and dangers. In agreement with the preamble of the European directive, about 60 % of the fatal accidents in construction could be avoided with the adoption of appropriate measures in the design or preparation phases [5]. There needs to be larger efforts to identify the risks in the design phase by architects and by engineers. The designers have a fundamental role in the choice of the safest options and they should consider the relative risks to the execution when they conceive the construction works. The construction owner and the designers should integrate safety's coordinator's activities along the elaboration of the designs.

This integration is justified for two reasons. The first reason, for questions of effectiveness, has to do with the safety's need to be integrated from the beginning of the creation of the design. This is justified because the decisions taken at the design level could lead to condition the effectiveness of prevention measures. The second reason, of intrinsic nature to the coordination activity, is linked with the fact that the activity of safety's coordination in the design phase should be coordinated with the other design specialities [6].

Construction Safety Coordinator Profile

Construction safety coordinator's in the design phase should be qualified technicians capable of articulating activities of designers and the construction owner's requirements [7]. This qualification should guarantee the technical, professional and personal competencies necessary to act effectively in these activities [8, 9]. Some of the competences, knowledge and aptitudes will be:

- To read and to interpret the several pieces of the construction designs;
- To coordinate prevention with designers and with the construction owner;
- To identify and prioritize the risks of accidents;
- To evaluate the current risks of the architectural solutions and adopted techniques;
- To present and to justify solutions to seek the prevention of professional risks;
- To understand the techniques and the constructive processes;
- To know how to apply techniques of administration of conflicts;
- To present and to justify, in the extent of the elaboration of the list of responsibilities, specifications that seek to prevent the risks of accidents;
- To analyze the proposals in the environment of the construction site to verify that they consecrate the prevention of accidents;
- To esteem the inherent costs of prevention in execution of the work.

These competences for the qualification of professionals in the area of construction safety have been defined by national agencies and by professional organizations. At the European level ISHCCO (International Safety and Health Construction Coordinators Organization) has defined a framework accepted by the national professional organizations of construction safety coordinators. This framework defines competences for levels 5, 6 and 7 of the European Qualification Framework in terms of knowledge, skills and attitudes. This framework intends to be a reference that will facilitate mobility of professionals and improvement of the quality of professionals working in the sector [10–12].

At another level the legislation imposes that the construction owner will also have to pay to assure that the design, and above all the construction, is carried out with safety. The lack of appropriate prevention measures will increase the costs that the construction owner, the contractor and the society will have to support. The costs will be, among others, repairing costs, production loss, loss of materials, medical treatments, legal procedures and increase of insurance fees [13].

Research Study with a Possible Solution

With the intention of answering some of the issues required to assist designers and construction owners directly, this research study was done with the aims of producing a model for integration of safety into the design process using a practical guide for designers. This analysis was based on the development of a risk assessment method for the design phase [14]. The envisaged model aimed at contributing to the prevention of risks of accidents in construction during the lifetime of the project (planning, implementation, maintenance and deconstruction), taking into consideration design decisions, accident risks and control measures. The research study consisted of the following steps:

(a) Identification of key stakeholders (owner, co-ordinator, designers, etc.) and their respective duties in construction safety, specifically in the sub-sector of buildings;
(b) Analysis of the design process;

(c) Search for statistics on construction accidents in order to understand the underlying causes and respective risks that originated those accidents;
(d) Analyse case studies in order to establish possible links between the causes of the accident and the design decisions;
(e) Method to assess risks at the design stage that could be eliminated or alleviated;
(f) Guide for designer containing guidelines for preventing accidents at the design phase.

Proposed Methods for Designers

The number and sources of accidents analysed was diversified in terms of sources. The accidents were obtained directly from public sources and from construction companies. This data from public sources was obtained from reports of accidents available to the public and from consultation of company records. This data obtained from the analysis of about 2000 fatal or serious accidents have shown that about 35 % of the accidents could have been avoided if, during the design phase, appropriate options had been taken. The same percentage was 30 % concerning decisions at the preconstruction phase, also known as the planning stage.

The research study also produced two models valuable for different types of designs (infrastructures, superstructures, mechanical, electrical, HVAC, architecture and plumbing): MAARD (Method of Analysis for Accident Related Design) and MMPtD (Management Model for Prevention through Design) [15].

MAARD is composed by a matrix that relates the frequency and the gravity of accident with the possible preventive measures to be considered at the phase of design. These preventive measures were chosen based on the risks that created the accident analysed. The measures were identified as possible to be decided during the design phase. This tool allowed the conclusion of how many accidents could have been prevented at the design phase, planning phase and construction phase.

MMPtD (Management Model for Prevention through Design) is composed of four sets of checklists that are supposed to be used by designers according to the respective type of design: architecture, structures, infrastructures and mechanical/electrical installations. These four guides are practical tools that can be used by any designer without an enlarged knowledge about prevention of accidents. It is expected that, if this guide is widely used by designers, there will be a serious reduction of accidents in construction, since more preventive measures will be undertaken at the design phase.

According to this research it is possible to include prevention measures that can reduce the risks that may create about two thirds of the accidents verified in about 2000 accidents. It is certain that the cause–effect relationship between the proposed measures and the accidents occurred may not be unique. There are probably other causes that could not be identified in that analysis of the reports that were not eliminated and the effectiveness of the proposed measures to eliminate the accident may be questionable. Nevertheless these are the best solutions according to the state of the art prevention culture. If these guides are followed by the designers in the design phase there will be a significant decrease of accidents in the related construction. One life saved is reason enough to apply these guides.

Safety in construction is a subject of life or death.

References

1. Construction Design and Management – CDM (2010) Available: http://www.hse.gov.uk. Access: 10 Jan 2015
2. Creaser W (2008) Prevention through Design (PtD) Safe design from an Australian perspective. J Safety Res 39:131–134
3. Driscoll T, Harrisson J, Bradley C, Newson R (2008) The role of design issues in work-related fatal injury in Australia. J Safety Res 39:209–214
4. European Agency for Safety and Health at Work – OSHA/EU – OSHA/EU (2009) Occupational safety and health and economic performance in small and medium-size enterprises: a review. Office for official publications of the European Communities, Luxembourg
5. European Foundation for the Improvement of Living and Working Conditions (1991) From drawing board to building site (EF/88/17/FR). European Foundation for the Improvement of Living and Working Conditions, Dublin
6. European Union (1992) Directiva 92/57/CEE do Conselho, de 24 de Junho de 1992. Available: http://europa.eu.in/eur-lex/. Access: 10 Feb 2014
7. Fernandes M, Soeiro A (2007) Analysis of the profile of competences of safety coordination. 7th international congress of safety and hygiene at work, Ordem dos Engenheiros, Porto
8. Gambatese J, Hinze J (1999) Addressing construction worker safety in the design phase: designing for construction worker safety. Autom Constr 8:643–649
9. Gambatese JA, Hinze J, Haas CT (1997) Tool to design for construction worker safety. J Archit Eng 3(1):32–41
10. ISHCCO (2015) International safety and health construction coordinators organization. www.ishcco.org. Access: 10 Jan 2015
11. National Institute for Occupational Safety and Health – NIOSH (2014) The National occupational research agenda. Available: www.cdc.gov/niosh/nora. Access: 30 May 2014
12. Oliveira OJ (2004) Gestão do processo de projecto na construção de edifícios. Integração, ano X(38): 201–217
13. Reis C, Soeiro A (2003) Economic analysis of implementation of safety plans in construction. J Kéramica 257:45–56
14. Soeiro A (2005) Safety in the construction. Edições FEUP. Ebook. Porto. ISBN 972-752-072-3
15. Vasconcelos B (2013) Segurança no Trabalho na Construção – Modelo de gestão de prevenção de acidentes para a fase de concepção. Doctoral (Ph.D.) thesis. Porto: Universidade do Porto

The Traditional Urban Square – A Vital Organ in the City or a "Thing" of the Past?

Fernando Brandão Alves

From the Definition to the Meaning of Square

In the strictest sense, a "square" is the result of a grouping of houses around a free space, whose meaning, implicit in the definition itself, stands out as its defining feature – it is a closed space.[1] Naturally, the definition of urban square raises discussion, given the divergent opinions on the role squares should play (or not) in the urban context today. However, from a more current definition of urban square – any space within the city, in which its defining elements are clearly visible from within and where the feeling of "being" is more acutely emphasised than in any other area – a number of features and essential conditions arise that need to be analysed[2]:

(a) Based on the definition put forward and the fact that the elements surrounding the square should be clearly visible, we can state that it is not the edified elements which delimit the square, but rather, that it is the square which bestows on them a tangential sense of boundary and conformation;
(b) Since the sense of "being" is characteristic, there has to be a certain level, even if minimal, of activity;
(c) The square's nature allows for maximum control over the space – easy external accessibility is associated with a minimum external surface to be controlled i.e., entries;
(d) Due to the square's affinity for the symbolic, it often displays public buildings or parts of these in its contents (*agora*, *forum*, church, cloister, mosque patio; etc.);
(e) The strong sense of containment or enclosure it transmits favours the development of certain potentials for animation (development of attractive activities, comfort, socialising, rest, among others).

Today, when we talk of the urban square, in its most classical or traditional conceptual sense, we tend toward a nostalgic feeling evoking memories of remarkable medieval or Renaissance squares erected throughout history and which, due to political or technical resolve in contemporary times, can still be seen, often peacefully relegated to an almost museological purpose on the city's tourist routes.

Even though squares or plazas are places where we may have experienced significant events in our lives, they are also points of reference in our orientation and appropriation of the environment around us. Today, more than ever, squares should be prepared to host a diversity of functions (places for demonstrations and socio-cultural, commercial or other types of gatherings (Fig. 1), places of rest and leisure, collective or individual, casual or programmed), capable of being remembered as a part of collective, personal or intimate experiences – the feeling of being is complemented by the development of one or more activities. The urban square cannot be drained of this vital function, especially when it also plays a role in rebalancing urban metabolisms feeding on phenomena of "ghettoisation", of real-estate opportunism, the blindness of a planning process which, at the beginning of the twenty-first century, is indoctrinated with the creed "what is new and different", another city which *Hall* described as "the city of the tarnished Belle Époque", the infocities or ghettos of misinformation of the recently inherited city.[3]

[1] The concept of square is associated with a notion of place – the interior that is experienced in contrast with a surrounding exterior.

[2] VIDE: BRANDÃO ALVES, F. [4] – "Avaliação da Qualidade do Espaço Publico Urbano. Proposta Metodológica", Fundação para a Ciência e Tecnologia/Fundação Calouste Gulbenkian, Lisboa.

[3] HALL, Peter [6] – "Cities of Tomorrow – An Intellectual History of Urban Planning and Design in the Twentieth Century", third edition, Chapter 12, p. 405, Blackwell Publishing, Oxford.

F.B. Alves (✉)
Faculty of Engineer, Oporto University, Porto, Portugal
e-mail: alves@fe.up.pt

V. Abrantes et al. (eds.), *The Pre-Fabrication of Building Facades*, Building Research: Design, Construction and Technologies, DOI 10.1007/978-3-319-22695-8_5

Fig. 1 Public space in new Forum, Barcelona (Source: archive of the author)

According to *Lynch,*[4] "*the plaza is intended as an activity focus, at the heart of some intensive area. Typically, it will be paved, enclosed by high-density structures, and surrounded by streets, or in contact with them. It contains features meant to attract groups of people and to facilitate meetings…*", clearly highlighting the notion of "containment" or "enclosure", one of the most relevant features of the square (and street). Its study provides a better understanding of the role of these morphological units within the city, highlighting particularly their mysterious ability to attract people, an important initial stimulus in the square's occupational dynamic.

The presence of the automobile in the square surely represents one of the best indicators of the nature and quality of these spaces, in which the pedestrian throughways, broad-walks, public pathways, among others, are also included. The square should comprise an open public space, contained by harmonious forms, interconnected with the other morphological urban elements (streets, other squares, broad-walks, among others), with paving throughout its main extension and where the continuous presence of private motorised vehicles could possibly be excluded (Fig. 2). More than a passageway, it should be a place in itself, whose main function is that of the interactive fulfilment of the human needs mentioned – such as taking a walk, sitting, contemplating, eating, reading, observing, talking and relaxing.

Fig. 2 Public space, Barcelona (Source: archive of the author)

[4]LYNCH, Kevin – *A Theory of good city form.* Cambridge, Mass: The MIT Press, [7], p. 443.

Recent literature on the analytical study, design and maintenance of the urban square is more focused on the development of universal axioms and methods to analyse forms than on exploring issues more directly related with the scale of the human dimension of public space, such as, for example, the spontaneous use of open spaces by regular citizens, or the behavioural aspects of the use of squares.

Note for example how rare the study and implementation of "full accessibility to public space" still is in the sphere of design and political decisions. *Marcus and Francis* are very sceptical when they state "*On the whole, those books that serve as inspirational and self-defining material for designers of the urban environment are theoretical (…).*

Certainly these larger issues are valid and important, but it is our fear that they often are the only inspirational source, to the detriment of the population affected by the resulting designed spaces".[5] Complementarily, other (rare) studies, particularly centred on pedestrian movements, are detailed documents whose main focus directed at the management, planning and supervision of urban areas, goes beyond the academism of design, in these cases, compensated by a number of ideas expressed in the selection of printed photographs:

"Streets for People", by the OECD,[6] highlighting the planning and management of urban space;

"Mennisker til fods" (People on foot), by Jan Gehl,[7] including studies by architecture students, on the behaviour of pedestrians in one of the oldest and most famous streets in Europe – Strøget – in Copenhagen, where, in the first year after it was transformed into a pedestrian throughway, the number of pedestrians rose by 35 %, and the number of baby strollers by 400 %;[8]

"Pedestrian Planning and Design", by John Fruin,[9] a statistical and detailed examination of the capacity for pedestrian circulation in streets, lifts, stairs, underground passageways, among others;

"Life Between Buildings: Using Public Space",[10] illustrating the multiplicity of open spaces that surround us, the daily activities and their specific demand in the environment created by man;

"Urban Space for Pedestrians", a report by Pushkarev and Zupan,[11] emphasising a sophisticated analysis of pedestrian behaviour in streets and squares;

"On Streets", a collection edited by Anderson,[12] on urban design and the social expression of streets, and which gathers important requirements formulated by the different authors on the design of squares and particularly on their interconnection with streets, from a joint perspective in which these and other leisure spaces are understood as essential places in the interactive use legitimately conferred on them;

"The Social Life of Small Urban Spaces", by Whyte,[13] focusing on a number of studies on squares in New York, with particular emphasis on the observation of behaviour in the environment as one of its main aims;

"Public Life in Urban Places: Architectural Characteristics Conducive to Public Life in European Cities" and "Livable Cities – People and Places: Social and Design Principles for the Future of the City", both by Crowhurst-Lennard,[14] which discuss the theory and practice of humanisation of the urban environment.

Does the Perfect Square Exist? How Is It Built?

An intrinsic feature of the square – "containment" or "enclosure" – has led to a variety of classifications according to the forms it can take. *Zucker*[15] distinguished five architectural forms: the closed square, where the space is self-contained; the dominated square, where the open space is directed towards a single structure or a group of important buildings, and all the other surrounding structures relate with it; the nuclear square, where the space is shaped around a centre; grouped squares, where the spatial units are associated in

[5] MARCUS, Clare Cooper; FRANCIS, Carolyn – *People Places – Design Guidelines for Urban Open Space*. New York: Van Nostrand Reinhold, [8], p. 10.

[6] Apud MARCUS, Clare Cooper (*et alt.*), cit. 5 (Reference OECD, Paris, 1974).

[7] Article published in the Danish journal Arkitekten, 70(2), 1968, pp. 429–446.

[8] GEHL, Jan – *Life between buildings: Using public space*. New York: Van Nostrand Reinhold, [5].

[9] Ap. MARCUS, Clare Cooper (*et alt.*), cit. 5. The authors mention the work of John J. Fruin – *Pedestrian Planning and Design*. New York: Metropolitan Association of Urban Designers and Environmental Planners, 1971, (original Danish publication, 1971).

[10] Ap. MARCUS, Clare Cooper (*et alt.*), cit. 5.

[11] Ap. MARCUS, Clare Cooper (*et alt.*), cit. 5. The work by Boris Pushkarev and Jeffrey Zupan is mentioned – *Urban Space for Pedestrians*. Cambridge, Mass: The MIT Press, 1975.

[12] ANDERSON, Stanford, ed. – *On Streets*. Cambridge, Mass: The MIT Press, [2].

[13] WHYTE, William H. – *The Social Life of Small Urban Spaces*. Washington, DC: Conservation Foundation, [11]. This work is the continuation of the study on several New York squares, in the scope of the Street Life Project in the 1970s, initially funded by the Rockefeller Foundation under the direction of William H. Whyte. Later, this Project was substituted by Project for Public Spaces, under the direction of the constancy firm Fred Kent III, which centred its studies on problem streets and squares in several cities.

[14] Ap. MARCUS, Clare Cooper (*et alt.*), cit. 5. Two volumes by Suzanne H. Crowhurst-Lennard and Henry L. Lennard are mentioned – *Public Life in Urban Spaces: social and architectural characteristics conducive to public life in European cities*. New York, Southampton: Gondolier Press, 1984; *Livable cities – People and Places: Social and Design Principles for the Future of the City*. New York, Southampton: Gondolier Press, 1987.

[15] ZUCKER, Paul – *Town and Square*. New York: Columbia University Press, [12], p. 151.

such a way as to form larger compositions; and the amorphous square, where the space is unlimited. For *Sitte,*[16] enclosure is taken as a prerequisite of the square, and he concluded that in formal terms there are only two types of square, identifiable by the nature of the dominate structure. Thus, he defined two categories: the "deep" type and the "wide" type, even though both may become apparent properties of a square when the observer stands at the extreme opposite to the largest building dominating the "*layout*". *Sitte* places greater emphasis on grouped squares than on the amorphous square or the inner space which surrounds the so traditional central element. He interpreted them not as generic forms but rather by the manner in which they are related to each other and to the urban fabric in general.

To achieve enclosure in the square necessarily implies analysing its corners. According to *Sitte,*[17] whenever possible, only one street should open out at a point, while a second street should branch off the previous one further back, out of view from the square. Overlapping views outward should be avoided from any point in the square. Traditionally, the edified structures are also a strategy in obtaining the sense of enclosure. Alternatively, the corner or entry can be closed, and instead an arch, a lintel or architecture itself can be used; see the case of the Arch of Rua Augusta in Lisbon which, due to its peculiar placement directly in line with the centre of the *Praça do Comércio* (Square of Commerce), establishes a transitional and articulated sculpturesque point from the street to the square, acquiring symbolic and referential qualities within this ample urban space.

Other, no less important, qualities of squares and their surrounding buildings affect their degree of enclosure. These include the nature of the buildings' eaves line or roofline, the relation between their height and the size of the space they enclose, their volumetry, the presence or absence of a unifying architectural theme, and the overall shape of the space itself.

In an internal space, the highest horizontal surface is usually the ceiling. Even though, by analogy, the heavenly dome may be the square's ceiling, it should in any case correspond to certain requirements in terms of metric composition.

For Zucker,[18] the height of the sky above an enclosed square should be imagined as corresponding to three or four times the height of the tallest building in the square; this relation seems to have more bearing particularly when the eaves or roofline is more or less of equal height to its length. Not infrequently, in many medieval squares, the variations in height are usually in the same magnitude of scale, where the picturesque nature of the rooflines or tie-beams stand out; enclosure is inversely proportional to the variation in height of the square's enclosing buildings.

The relationship between the effective height of the buildings and the width of the space is always a critical issue if a harmonious space is to be (re)created. If they are too high in relation to width, a feeling of oppression may arise; if they are too low, a strong sense of vulnerability and exposure. It could be suggested that the maximum harmonious proportion of height to width should be 1:4, respectively; that is, a comfortable proportion so that an observer at the centre of the space can truly experience it from every angle. The numbers put forward are undoubtedly less modest than those of *Alberti* or *Palladio.*[19] According to the former, the acceptable variation is between a third (maximum) and a sixth (minimum) in height in relation to width. Palladio, though, narrows *Alberti's* proportions, such that the square's width varies between 7/4 $\left(1\frac{3}{4}\right)$ and 5/2 $\left(2\frac{1}{2}\right)$ times the height of the buildings, respectively, based on the typical width of the Roman Forum.

Based on experiments carried out in the field, we believe that the detail of a building is better perceived at a distance equal to the building's largest dimension. On the one hand, some theorisers contend that the building is best seen as a whole, i.e., as a total composition at a distance approximately equal to double its height, or at the distance given by a projection of a line on the ground at an angle of 27° drawn from the eaves line. It is believed to be the most comfortable proportion so that an observer in the centre of the space can truly experience it from every angle. To take in more than one building requires a distance from the observer of three times their height or a distance which follows the previous calculations at an angle of 18°. Below this measure, the objects lose predominance in the field of vision – other objects beyond the square can thus be perceived and the sense of enclosure is lost.

Sitte stated that once the height of the main buildings was taken, the square's minimum width and maximum dimension could be declared, so as to obtain the most favourable perception, as being treble the height (a proportion of 3:1). Furthermore, the general form of the building, its purpose and detailing could not admit exceptional dimensions. Only with this metric proportion would it be possible to truly enjoy the entire physical and perspective dynamic of the space, bearing in mind the physiological limitations of human sight and the full range of sensations it provides. Despite all these principles, there are many a successful square which do not obey any of these normative restrictions. It should be noted though that *Sitte* was highly influenced by the small-scale medieval square. Some squares in these conditions may have merit all of their own, whether due to their absolute dimensions, even if reduced, or their symbolic value

[16] MOUGHTIN, Cliff – *Urban Design: Street and Square*. Oxford: Butterworth Architecture, [9], p. 99.

[17] Idem-Ibidem, p 99.

[18] Ap. MOUGHTIN, Cliff, cit. 16, p. 99.

[19] Idem-Ibidem, pp. 100–101.

for the community, or for other reasons. That in which they may fail, in terms of the sense of enclosure, they usually make up for in their sense of place and not least in the vibrant activities they support.

The absolute size of urban space is also related with its degree of enclosure or containment. *Sitte* found that the largest squares of old cities were on average only 57 m×143 m. Many of the most charming squares, in the historical areas of our cities, are as small as 15–21 m, which today is barely wide enough for a road.

The restrictions imposed by human optical geometry, at the scale of the city, indicate that the limit to distinguish human gestures is about 135 m. Obviously, the distance at which someone can perceive other movements – military parades, fireworks, etc., is much greater than that required for human gestures. Let us admit that an observer at the centre of a space can turn around and take in every side of that space if the height-width proportion is 4:1. Even though the relation between buildings and squares can be established as definitive, the (hypothetical) metrics for visual balance which were mentioned previously may provide a guideline not only for certain interventions, but also to refine the critical perception of the spaces we observe. Thus, a square with 3-storey buildings should be about 36–45 m in width, and those with 4-storey buildings should be 48–54 m. If however the aim is to perceive the entire composition of the square's façades or of a group of buildings, the distance should be treble the height.

The maximum of 135 m mentioned for size admits the existence of buildings with about seven storeys, i.e., the movement of the observer in the square provides him/her with a reading of the composition as a whole, of the buildings' proportions individually, and also of details when focusing more closely, although the sense of enclosure is attenuated. The higher the perception of the tri-dimensional modelling of the surrounding buildings, the more reduced the sense of enclosure of the public space. Containment or enclosure is lost if, for example, the space's boundaries are shaped by town blocks or isolated neighbourhoods. For example, *Rob Krier* has in many cases opted for solutions employing the open design of façades for urban space, in an attempt to find other purposes other than enclosure. Repeatedly, the ideal of enclosure is the bi-dimensional quality of the plan.

The buildings around an enclosed space should form a continuous surface and seem an architectural unit to the beholder (Fig. 3). This property can be clearly perceived in the use of colonnades and arcades as continuous forms of connecting the ground floors of different buildings, creating a gallery or covered passageway. Some theorisers contend that the ideal distance to clearly perceive a dominant building at the extreme end of a square is approximately double the building's height, measured perpendicularly to its main façade.[20] Even if these metric relations have proven fundamental in the reading of the detail of Gothic buildings, including their statues and sculpture, their importance is often best understood in the oldest religious squares of medieval times.[21]

Fig. 3 Venceslau Square, Prague (Source: archive of the author)

A remarkable Renaissance example is the *Piazza Della Santíssima Annunziata* (mid-fourteenth century), which took on the name of the Basilica. Small, rectangular and welcoming, the square closes off on one side the large axis of the present-day *Via del Servi*, which in turn is closed off at the other end by *Brunelleschi's* great "dome". One of the features which most contributes to its charm are the three lateral galleries lining its boundaries. In general terms, its current

[20] The square which contains all these rules is *Piazza Navona,* in Rome, whose sides maintain a relation of 1:5 approximately.

[21] Sitte revealed preferential relations between length and width but did not neglect to mention that in great squares where this relation is greater than 1:3, the space loses part of its charm. For Alberti, the ideal of the square is centred on a relation in which the length is double the width.

appearance represents a concentration of urban planning as envisaged by *Brunelleschi* and his contemporaries. Between the thirteen and seventeenth centuries, several artists were involved in the design of this part of Florence, following the imperatives of the place as highlighted by their predecessors. If the *genius locus* has become a lost art in recent times, undeniably essential in the grand construction of the city, in Florence this art was demonstrated by all those who contributed to the creation of this square.

The Square as a "Dominated" or "Enclosed" Form

The first two of *Zucher's* categories – the "enclosed" square and the "dominated" square (the latter equivalent to *Sitte's* "deep" and "wide" squares, mentioned earlier) – are nothing less than variations of the same type, distinguished by a quality which they share – "enclosure" or "containment" – the purest expression of a sense of place.

According to *Zucher,*[22] the dominated square is characterised by the presence of a singular structure or a group of buildings with which the open space establishes a direct relationship and with which all the other surrounding structures are also related. We have seen that, for *Sitte*, the classification of squares was restricted to only the "*deep*" and "*wide*" types; both fall within *Zucher's* "dominated" square category, in which being deep or wide usually becomes an apparent property whenever the observer is opposite the main building which dominates the entire space.

Indeed, the building dominating a deep square should have dimensions which are proportional to the space it is directed at; in the past, it was usually the façade which complied with this requirement. The medieval space in front of the building constituted an extension to the function of its main entrance – here the religious community would gather, before and after the service, an example of which is the traditional churchyard, where sermons were preached outdoors and from which great processions departed. The buildings around the church where almost always related with these functions and were, naturally, subordinate to the main structure.

The Square as a Belvedere

A public square can be dominated by a view or not infrequently by a building or group of buildings or objects of great sculptural value. In several squares or plazas in southern Italy or in Sicily, the space is shaped by buildings which only line three of their sides. The fourth side is a belvedere which permits magnificent views of the landscape lying beyond the square. See the case of the *Praça do Comércio* in Lisbon, by Eugénio dos Santos, facing the Tagus River, which opens out into a striking view of the water's surface, making it perhaps one of the most impressive squares in the Iberian Peninsula. In the ideological fashion of the Enlightenment of the government of the Marquis de Pombal, it constitutes an "(…) admirable open stage over the Tagus",[23] which drops the curtain on the Reconstruction plan (following the 1765 earthquake that ruined most of the city) of this part of Lisbon. We can find other examples of belvedere squares or plazas, particularly in cities built on sloping waterfronts, such as the notable case of the city of Taormina in Sicily.

The Square as a Point of Departure and Arrival

As a point of departure and arrival, the square gains its greatest expression in the *Piazza del Campidoglio*, in Rome. It was during the papacy of Paul III in 1537 that Michelangelo was commissioned to project a monumental plaza on the *Campidoglio* hill; it was only concluded about a hundred years after his death. Here, thanks to the attention paid to architectural detail, the author created a unifying composition dominated by the direction defined by the main building, the *Palazzo del Senatore* and, in the opposite direction, by the views over Rome. The design was also restricted by another existing feature, the *Palazzo dei Conservatori*. Michelangelo proposed new architectural forms for both palaces in the mid-sixteenth century.

This intervention in *Campidoglio* illustrates the communion between the first Renaissance squares, such as the *Piazza Della Santíssima Annunziata*, in Florence, and the interventions in Rome in the late Baroque period. Despite the trapezoidal geometry of the square, the design of a starburst pattern in the pavement, at the centre of which stands the statue of Marcus Aurelius, spreading over the oval courtyard, bestows the square an illusion of rectangularity – the well-known effect of false perspective which so profoundly marked the works of this Master, and which is brought on by the narrow alignment of the existing buildings. It is furthermore an extremely exquisite example of the "radiance effect" in urban space, which among other particularities, conditions and directs the movement of pedestrians and brings on the optical illusion. In summary, the success of the intervention is the result of an urban design which, in light of pre-existing constraints, acquired the necessary creative and evocative

[22] Ap. ZUCKER, Paul cit. 15.

[23] AUGUSTO FRANÇA, José – *Lisboa: urbanismo e arquitectura*. Lisboa: Instituto de Cultura e Língua Portuguesa, (1st edition from 1980), [3], p. 46.

Fig. 4 Public space in the water front area of Cape Town (Source: archive of the author)

value without destroying its cultural legacy. On the contrary, the symbolic value of its essence was only enhanced.

Naturally, other spaces, roundabouts, some forms of broad-walks or, recalling Zucker, the "amorphous" or "nuclear" square, cannot be included in the category of public square, given their characteristics; notwithstanding the importance they may have as spatial features, their design requires considerations which differ from the types analysed above.

The Square as Centre

Undoubtedly one of the most important elements in urban design, the square has been one of the spaces most sought after for the location of public and commercial buildings in cities. It is at the same time an area enclosed by buildings and an area designed to exhibit buildings in all their splendour. Great compositions such as the *Piazza San Marco* in Venice, the *Piazza San Pietro* in Rome, and the group of squares in *Bath* by *John Wood* (and son), are unique in the qualities regarding spatial organisation, surrounding buildings and the plasticity of the silhouettes of their roofing; they achieve a strong emotional meaning and, as such, are comparable to any other form of art.

The activity of the square is important for its vitality and also for is visual attractiveness (Fig. 4).

On the design of the Roman Forum, *Vitruvius*[24] said that these aspects should be proportional to the number of inhabitants, such that it should not be too small to be useful or that it seemed excessive.

The Renaissance theorisers followed these principles. *Alberti*[25] added that there should be several squares throughout the city, some to give place to commercial activities in times of peace, others dedicated to activities proper of youth, and others still to store provisions in times of war. He went as far as to detail several types of mercantile squares, some for gold and silver, others for spices, those for wood, and those for livestock, etc., examples of which are *Praça das Flores*, in Porto, *Praça* (or *Largo) do Toural*, in Guimarães, among many others; each should bear appropriate detailing and occupy a specific place in the city.

However, to transfer concepts or principles of urban design, which were once useful in certain places, to new realities, may represent some risk. The great virtue of the wonderful squares or plazas of Italy can in part be explained by the combination of climatic conditions which encourage life outdoors and the temperamental dispositions which characterise Italian culture. These conditions and the spontaneity of the Mediterranean populations stimulate public life which in itself bestows form on the square and street.

The mono-functionalist practices, of separating and segregating functions, associated with the architecture and urban planning of the Modern Movement, were shown to be a drawback in the art of building a city. The product of this line of thought, the massive complexes of services buildings or the large commercial precincts, have immobilised large areas of the city by closing their activities at the end of the

[24] VITRUVIUS - *The Ten Books of Architecture*. New York: Dover Publications, [10]. Book V, Chapter 1, p. 132 (trad. por Morris H. Morgan).

[25] ALBERTI, Leon Battista - *The Ten Books of Architecture* (1755 Leoni edn). New York: Dover Publications, [1]. Book IV, Chapter VIII, p. 81.

day. The most successful urban squares, even though they possess a dominant function for which they are known and by which they are classified, are in most cases those that ensure a strong dynamic by the diversity of usage day and night.

The singular most important function of an element within a city is its underlying symbolism. The greatest manifestations of art are intimately linked with our deepest feelings and emotions. According to *Moughtin,*[26] the great square is also linked to the world of fantasy, to the context of feeling. This primitive reaction to the world around us, including the edified environment, is intimately and undoubtedly linked to the way in which we, also, understand the human body – a type of "standard building" of urban design.

Human perception of space is centred in each of us. The development of schemes of spatial organisation based on this subjective idea of centre is extended to the notion of external centre[27] as a reference value in the environment. This idea is applicable both to the known world of each individual's daily life and to the external world, hostile and undifferentiated. As an extreme opposite to the public concept of *World Centre*, we find the house or family as put forward by *Norberg-Schulz,*[28] when he argues that, if the centre of a world designates an ideal, a public objective or "paradise lost", the "house" world possess a much stronger and embracing concrete meaning which, in other words, means that each individual possesses a centre to his/her proper world. In this interval of extremes – world and house – there is a continuity of hierarchal centres which serve different communities and which are the underpinning of the disciplines of urban architecture, design and planning.[29]

The centre is dominant in the city, distinguished from other places. It is only when one reaches the main square of many of the old cities that there is a feeling of having truly "arrived"; all streets natural lead to this focal point. A few European cities have maintained the importance of their centre – *Market Square* or *Slab Square,* as it is affectionately known by the inhabitants of Nottingham, is still today an opportunity for social life and the centre of many and diversified activities.

Among the non-green spaces, the urban square is unequalled as the best equipped space of reception or stay in the city, as proven by the *Piazza San Pietro* in Rome, completed by *Bernini* between 1656 and 1667, an important reference point in Rome's urban structure and, at the same time, the geographic centre of the Catholic world.

The Square as Gateway to the City

Any place has the dual function of entry and exit. It becomes a centre because it constitutes an objective; a place of pilgrimage, of popular demonstration, or often a place to supply the population, etc. In the same way, the function of "point of departure" or "point of arrival" is also significant (Fig. 5). This tension between centripetal and centrifugal forces is more visible in the portico, so clearly explored by *Alberti*[30] as the objective part of the city where the beginning of a trip is defined or, on the contrary, the place where one arrives and defines a new period of rest.

From Antiquity, the gateway has played an important role in urban and architectural design. See the example of the *Piazza del Popolo* in Rome which, for centuries, until the age of the railway, constituted the main entry to and exit from Rome for all the visitors coming from the North or those who departed in that direction.

We can take the gateway to mean an "invitation" or a "barrier". The transition from one domain to the other is also a critical issue in the design of the city's organisation (Fig. 6); more than in the definition of *Alberti*, and without forgetting the importance of redefining entrances and exits at certain strategic points, not only for the city but also for its most relevant spaces, to mark a transition should mean greater fluidity and less hesitation in entering or exiting any delimited area. Today, the entry function is different in the urban fabric; however, its function continues to be present in certain areas. The proposals designed should offer subtle creative solutions, in which the main concern is the organising effect of the entire composition of the space and not resort to hostile physical elements.

Final Note

The reflection on the meaning of place, as well as the citizen's connection with it and in particular with public space, may constitute an aid in better understanding the needs and rights of citizens in public space, in terms of its human dimensions. Given the growing migratory phenomenon between countries and cities, where diversity and the confrontation between ethnic communities is increasingly greater, notwithstanding diplomatic restrictions in some cases, and the growing free circulation of people and goods

[26] MOUGHTIN, Cliff, cit. 33, pp. 88–89.

[27] MOUGHTIN, Cliff, cit. 33, p. 89.

[28] NORBERG-SCHULZ, Christian, cit. 37, p. 19.

[29] According to Christopher Alexander, the whole should be a "centre" in itself and should also produce a system of centres around it; the centre tends towards symmetry, particularly bilateral symmetry, similar to that of the human body. The formation of the centre takes on the profile of a natural object, self-determining. This magical relationship between the centre and the complexity of the surrounding urban space comprises a unifying potential in the constitution of the whole. At that time, the plan and the project very simply work as a set of natural forces.

[30] ALBERTI, Leon Battista, cit. 77, Book IV. Chapter VIII, p. 80.

Fig. 5 Old centre, Guimarães (Source: archive of the author)

Fig. 6 The sense of entrance/passage, Barcelona (Source: archive of the author)

in others, the meaning of place and consequently of man's connection with the environment are increasingly urgent and comprise fundamental aims in achieving and preserving the quality of the urban environment; this occurs particularly when people grow roots in a certain areas, in such a way that they become important elements in their lives.

It is the spaces which we inhabit and experience, and the activities that there take place, that should sustain man's connections to a place, since they are undoubtedly a primary need for any individual; people's interaction with the place, individually or as a group, encompasses both their connection with the historical, socio-cultural, economic and political dimensions, and the symbolic spectrum of their connection with the Universe, or other worlds, where their biological and psychological nature bears weight, as does intellectual development, education, and sexuality, in a complexity where time and space express themselves simultaneously and on equal terms. In this context, and recalling in this article all the architectural and urban qualities which characterise (and should continue to do so) the urban square, there are no doubts as to their potential as a vital organ in the contemporary urban metabolism, as well as the importance of their role in bolstering the city's social cohesion and, as such, the quality of life for all citizens.

References

1. Alberti LB (1986) The ten books of architecture, 1755 Leoni edn. Dover Publications, Book IV, New York
2. Anderson S (ed) (1978) On streets. The MIT Press, Cambridge, MA
3. Augusto França J (1989) Lisboa: urbanismo e arquitectura. Instituto de Cultura e Língua Portuguesa (1ª edição de 1980), Lisboa
4. Brandão Alves F (2003) "Avaliação da Qualidade do Espaço Público Urbano. Proposta Metodológica", Fundação para a Ciência e Tecnologia/Fundação Calouste Gulbenkian, Lisboa

5. Gehl J (1987) Life between buildings: using public space. Van Nostrand Reinhold, New York
6. Hall P (2002) Cities of tomorrow – an intellectual history of urban planning and design in the twentieth century, 3rd edn, Chapter 12. Blackwell Publishing, Oxford, p 405
7. Lynch K (1981) A theory of good city form. The MIT Press, Cambridge, MA
8. Marcus CC, Francis C (1990) People places – design guidelines for urban open space. Van Nostrand Reinhold, New York
9. Moughtin C (1992) Urban design: street and square. Butterworth Architecture, Oxford
10. Vitruvius (1960) The ten books of architecture. Dover Publications, Book V (trad. por Morris H. Morgan), New York
11. Whyte WH (1980) The social life of small urban spaces. Conservation Foundation, Washington, DC
12. Zucker P (1959) Town and square. Columbia University Press, New York

Part II

Northwest Science Building of the University of Columbia Building

The Structural Diagram to Define the Facades Design

Construction to Discipline Architecture

Conversation with Rafael Moneo on Northwest Science Building of the University of Columbia Building about Technology and Architecture Methodology

Bárbara Rangel, José Manuel Amorim Faria, and João Poças Martins

In 2011, for the third edition of CdO, we interviewed Rafael Moneo to understand the concept the ruled the design of the Columbia University newest building. For the extension of the Faculty of Sciences at Columbia University, Rafael Moneo, besides the responsibility to complete the Columbia Quarter in the Manhattan grid, had a difficult task, constructing a building over a gymnasium with a span of over 40 m. The solution was to build a bridge building over the void left by the gymnasium, compensated with a wind bracing system in the façade. To support this "bridge building", the façade had to work together with the main structure, through a diagonal bracing system. This apparent difficulty became the embryo of the design concept. The structural engineer defined the façade structural bracing system and the architecture used that matrix to design the constructive system to compose the façade. This mathematical design became the formula to solve the entire program, as Rafael Moneo confesses:

> *In a sense I had not expected such a powerful and conspicuous structure, or that we might make of that structure the virtual substance of NWSB's architecture. The structure reflects all the nuances that the singular spatial organization of the building requests*[1]

This mathematical formula to solve the façade was also the answer to the urban challenge to complete one of the most important blocks on the orthogonal grid of Manhattan. To solve this project, Rafael Moneo followed the methodology of the development of the city itself, a city that is built on mathematical rigour on every scale, imposed by the construction in solid brick, through the consequent modulation of the façade and the block, to the strictly orthogonal metric of the city of New York, broken by the exception that is the Broadway.

Like Mies van der Rohe when arriving in the United States, he understood how the constructive material of the composition of the building, specifically the structure, could also be a tool for the composition of the architectural language. In this project, Rafael Moneo moved a step further. The design composition of the façade is achieved through an iterative process based on the structural model. Even the configuration of the covering steel constructive system panels respects this design methodology.

Alongside the first moments of conception in the architectural studio, the structural team developed a computer model that defined the importance of each of the diagonals in the stability of the façade of the building, showing those that worked in compression and traction. This graphic was gradually fine-tuned, removing the diagonals principally in compression, therefore resulting in a graphic schematic showing the movement of the forces on the façade: The base schematic for the architectural design.

> *I like to think that we have achieved an integrated manner of relating all the means of architectural production, from the conception of the volume, to its structural engineering, to the detailing, as well as the issues of construction.*

With this interview we tried to reveal the process of this design methodology bound in the complicit articulation of architecture and engineering.

[1]Moneo, José Rafael and Brock, Jeffrey. *Northwest Sciences Building*, for "Post-Ductility: Metals in Architecture and Engineering" Princeton Architecture Press, 2012; Cadernos d'Obra 03, Porto, May 2011

B. Rangel (✉) • J.M.A. Faria
Faculty of Engineer, Oporto University, Porto, Portugal
e-mail: brangel@fe.up.pt; jmfaria@fe.up.pt; jppm@fe.up.pt

J.P. Martins
Civil Engineering, University of Porto, Porto, Portugal

V. Abrantes et al. (eds.), *The Pre-Fabrication of Building Facades*, Building Research: Design, Construction and Technologies,
DOI 10.1007/978-3-319-22695-8_6

Designing for a New York Funded Block

How is your design methodology articulated with the engineer's? This project is a paradigmatic example of your design methodology, a constructive solution to solve an urban problem and functional program. Here that relation is driven to the extreme, the structural solution was the architectural design concept. This was an inversion of the design methodology between architecture and engineering. The rules were first defined by the structure and the architecture explored the design solutions according to that matrix. How was this inverted process developed?

I think that this project, more than any other or as much as any other, obliged us to solve constructive problems from the word go.

From a general point of view, the project was already given, it had to occupy the last building site on the Columbia campus. What happened is that the lower storey of the plot was already taken up by a gymnasium. It was necessary to build over the gym to bridge a 40 m span. To bridge this span could almost be said to be the first baseline for the project. As I say this, I realise that it is only partially the case, because it must be said that the way in which the building would bridge the gym was pre-ordained by the massing conditions of the design and these massing conditions did not refer so directly to having to bridge the gym, but respected the massing provisions and the urban planning conditions of the McKim, Mead & White Columbia campus.

It must certainly be said that this was the first urban planning vision of the project, a vision that led us to think that the best to be done in building on the site was to serve and abide by the norms established by the campus designers who had proved so efficient in leaving it embedded in a city like New York.

So let's admit that, following this first idea of how to erect a building serving the massing established by McKim, Mead & White, we found ourselves obliged to bridge the gym. Since we must bridge the gym, the issue of how to resolve the problem structurally became capital, crucial. And it's at this point that the idea, not new in fact, of making the façades, the vertical planes that define the massing, to be those that enabled the bridging of the gym, became the key theme of the design. So, the construction of this prism that bridges that void and gives rise to a structure that is taken as the existing frame of reference with which the architect must grapple and struggle, was at the root of the project.

You always speak in your texts about the site, saying that Nature is born out of the expectant site and the restlessness of the architect. What were Stirling's and your own attitudes to the site? Did you bear in mind Stirling's work for the same plot?

We should say that Stirling's design was a design that considered that that corner must be resolved with a singular building, while ours wished to reduce this singularity, forget it, and give primacy and preference to the weight of the campus structure taken as a whole.

In principle, the attitude was the opposite of that taken by Stirling. In Stirling's design the building was, shall we say, the main player. In our case, we put the emphasis on the campus as a whole. The first strategy was to think of the building as another piece in the campus and not as an isolated building. Stirling's building introduced the singularity of that slanted direction which was, on the other hand, an existing virtual guideline, because the contact between the lower level of Broadway 120 and the campus made this slanting mandatory. But this slant also introduced a geometric component that was completely different from the orthogonal order of McKim, Mead & White, which is respected in our massing. All this also happened outside strictly linguistic issues. Stirling's design dates back to 1984, I think, a design quite marked by the stylistic problems that were being debated at the time; it is clearly a post-modern design, so to simplify the question you asked and reply with a straightforward yes or no, if Stirling's building influenced our proposal, I should clearly say no.

Looking into these two solutions to the same project, yours and Stirling's, how does the architectural arbitrariness come into the design, in a brief and a site with so many limitations such as the existing urban plan and an urban image that was already very strong?

I think this is a design… well, this also happens in many others, in which the presence of contingent elements that force a choice for the specific appears very clearly. Where I think that a more individual vision of this architecture emerges – I'm loathe to use the word personal – is perhaps in the fact that it's a building that in its first physical erection emerges as a simple prismatic mass and, in the end, once it accepts all external inputs, acquires a tremendous complexity. The massing simplicity of the design does not tally with this dense and diverse condition that is apparent in many little occurrences and in many small architectural episodes, that can actually be considered separately, but which in the end are packed into a compact, close-knit mass which is the massing that we're talking about. This is quite visible in the section.

The section shows how this closed box – which then vibrates given the structure of the closure – is nevertheless occupied by an overlap of uses, functions and movements. Basically, it's an intricate, complex building, despite its apparent simplicity. Buildings are often deceptive, or confusing as to what they really contain. This seems to me to be the case with Columbia.

In this project, the programme is complex and the situation of the surroundings very complicated. What was the solution?

The surroundings are complicated and also rather variable. The laboratories, practically, were built on nine levels, each with very different characteristics. In fact no one laboratory is the same as another, simply because in the same building there are physics, chemistry and biology labs, and teachers with syllabuses or agendas that require different uses. Only the basic structure of these spaces – that provides for their use as laboratories – and the double height which teachers and students share, is the same. But the life of each lab should be seen as independent. It is in all these shared spaces (libraries, classrooms, cafés, lounges and meeting rooms, lecture halls) that the diversity we're talking about manifests itself. The other important thing that it shows is that it is very much a university building, but one that is also very accessible to people. Basically, the Columbia campus has this attraction of being and not being in the city. This building is a permeable, open building, it's a point at which city life and the life of the students mingle. This also makes the building attractive, makes it responsible for that complexity we were talking about.

Pursuing this town planning issue, how has it been possible to make this volume so different from the neoclassical context? Was the height of the building defined by the brief?

Actually, I think the constraints in New York are less stringent than we sometimes have in Europe. I think we could have gone even higher, it was possible, but the University restricted the use of land so as to avoid misgivings on the part of the community. The neighbours must be heard on those issues, their opinion matters. Actually, during the process, we had two or three public hearings at City Hall. According to professional practice in the United States, this issue didn't fall on my shoulders, it was managed more independently by the University and the City Hall without making the architect responsible for its management. In this, work practices are different in the US and in Europe.

Did the selection of the material for the façade involve a difficult negotiation?

The most important problem was more focused on the material. People said that they understood Columbia as a university built in brick. And it's true, it was, originally, but when you look at exactly how it was built, you see that there are other materials besides brick. But brick prevails as the material with which the campus was built. And in fact, at some point in the project, the argument arose for not using brick.

For me, and it's a material I've used a lot, brick is increasingly hard to use, and it's particularly difficult in countries where the bricklayer's craft has disappeared. Admittedly, much of the History of Architecture has been achieved with the help of what we call craftsmen, and when this craftsmanship disappears, those materials and systems of construction that have been extraordinarily sound are manipulated in such a way that they become distorted. I hate it when I sometimes see someone using bricks to erect prefabs, for example. The brick is doubtless associated with a way of making that is reluctant to admit the current construction uses and techniques. That's why I resist it. Apart from that, and although it's true that you can find brick buildings of 20, 22 storeys like this one – there are many in Manhattan – brick does seem to be a more appropriate material for buildings of no great height.

It would have been impossible, actually, to construct the building in brick. On the other hand, if something is asked of a material, it is that it is harmonised, not contradictory, with the uses of the building. So we wanted a material that could be seen as more sophisticated. Aluminium is closest to the idea of how we understand today the image of the industry. I don't know, aluminium is still used on aircraft…

Might it also be that the choice of construction system and the image of metal could mean a break that the urban response has not raised in the project? The urban response has been very respectful of the surroundings…

Yes, it's true, that's the licence one takes. Although I'm not saying that on occasion one can't be allowed to build almost replicating an old building. For example, the case of the corner of the Bank of Spain seemed to me an intervention so small, so minimal, that it made no sense to ignore the more unified vision of the building. But usually the opposite happens, that a more natural expression leads us to build without the prejudice of having to follow a form or a language that we're not used to today. I also think that the success of the McKim, Mead & White campus has been in keeping the same idea of urban planning strategy for a hundred years without submitting to what architecture was at the time the project was proposed, in the late 19th, early 20th Centuries.

Each of your works contains a deep theoretical reflection. Murcia, the Kursaal and Columbia are three very different responses to three very different environments. Murcia in a very classical plaza and with a very strong weight of urban history, the Kursaal is an intervention at the meeting point of a city with the sea and Columbia is located in a very different urban setting from the others. How do you see this relationship?

They are three very different buildings. If we were talking about the more superficial, epithelial aspect, of the façade, we would find more points of contact between Murcia and Columbia than between Columbia and the Kursaal. The Kursaal, after all, is only a volumetric building and, ultimately, what is required there is to find out what materials to build that volume with, which one wants to make very abstract but must lose some of that abstractness in the building of it. But the building is complete in the volume. The textural problems the Kursaal has are that it has very little thickness, there is hardly any bulk, and it is practically unseen. What remains is just the volume, the mass of the building. In the case of Columbia, it's a problem of language, texture and structure. The difference between Murcia and Columbia is that in Murcia the altarpiece side of the building is dictated by purely visual values.

Here, what has been attempted is to try to match the visual values with the structural. In a way, one could say that in a building like Columbia there is a certain understanding of the architecture from the picturesque, from the diverse and the varied or an acceptance of what is the structure dictated by necessity, because the structure is the starting point. This project began with the solution provided by the structural engineer, once we had decided that the block spanned the gymnasium. That it spanned the gym and other things too because, of course, this block bridges the gym but has other problems. If you look here [pointing at the plans] this is the line we couldn't pass, because it was at this point, here, where the tracks were, this was a crucial point for the vertical communications. We couldn't have vertical communications and we also wanted the building to be very permeable from there. Therefore, the building retrieves all this light, but there are also wrought elements hanging from the structure that make it possible to access the building from the street as if the whole volume were suspended, in flight. To simplify it, it can be said that the façade resolves the bridging, but besides solving this problem there are other elements and other demands that make the project more complex. And that complexity is accepted and is transformed into something that is also "admissible" visually, but in reality it's not only admissible, but fulfils the expectations that I had in visual terms for the building. This makes it a little different. It's true that in both Murcia and Columbia there's a certain randomness and disorder in the structure that, in the case of Murcia, only the architect manipulates and dictates and, in the case of Columbia, is also dictated by the engineer and the resistant structure.

The Structure as the Rule of a Numerical Design

Here, in both Columbia and Murcia, you play with the ambiguity between structure and ornament. Was the structure of the façade in this building composed at once by the architect and the engineer…? Because in Murcia it stems from the altarpiece, right?

In Murcia it only came from the architect. In Murcia the architect provided the engineer with what to do and here it was the engineer who established the first patterns for the geometry of the building. What happens is that, really, it's in the manipulation of those guidelines by the architect that his work is focused.

At this time I'd also say that without architectural intervention the engineer's guidelines have no value. What this means is the response to a more general problem of the History of Architecture, which refers to the manner in which a structure is filled in. From our western culture, since gothic architecture, the difference between what architecture and infilling is has been a generic and substantial question. In this case it's a hybrid solution and, in a way it is not ambiguous. It's more hybrid than ambiguous. Because it's not like in some buildings by Mies, in which the infilling works directly with the structure. The structure here is behind, and is occupying exactly all the guidelines for the finish of the aluminum façade. But it's in this adjustment, that which goes from the infilling to the structure, that the work of the architect lies, which in this case also entails – for programmatic reasons too – that for the labs it somehow makes no difference where the windows are and where they're not. Hence the randomness, or rather the need for the structure to become random almost with an aesthetic brand. This is the question that arises in a project like this.

In this case does the rule of composition come from the structure? What rules were imposed by the structure?

Well, here there's not so much a numeric component, though it exists … The question you ask is pertinent because in this project the condition of overlapping horizontal planes is what provides the key visual structure which then … let's see how to explain it…

In fact the building is these horizontal planes, but then what you are working is the façade, is the vertical plane. The numeric appears more from the geometry than simply from the numbers. In this sense, this façade is more geometric than that of Murcia, which would be more numerical. But that's also linked to the constraint of having to accept here that the laboratory is the overlap of these horizontal planes. From these you take the problem to the vertical plane and the façade and, therefore, the geometry, and not only the numerical order.

The organisation that you made of the brief, almost typifying the laboratories and offices, has left you free to compose the front almost like an abstract drawing, as an abstract composition, in direct response to the program.

We shouldn't talk of "composing" the façade, because this façade is not "composed". In this sense, in fact, the work of the architect is more in the manipulation of this surface geometry, in trying to transform the structure dictated by necessity into a consistent visual substance. It may also be a game on our part. The engineer wouldn't have cared if we introduced other diagonals for composition, but we haven't.

You could say that the rules were made by the engineer.

Yes, well, and accepted, rather, used by us. Not that there is no work by the architect, it's just that there's none of the architect's work in the composition. You could probably use this term in the sense of music. This project is more reminiscent of the studies a composer makes for a fugue than that of someone aiming to shape a melody. And it could also be said that Murcia, in that sense, is more concerned that the façade has something melodic, while here is more a reference to the fugue, to music that responds to itself, where what you find are the echoes of the same problem on another higher level, but always the same problem, the same issue.

It makes sense to speak of Murcia, but I understand that what you have to address is the differences between Murcia and Columbia. It's these differences that matter, not the similarities.

This building is derived from a structure, a façade that is separable, which is restricted by the structure. But what was the reason to suspend the floors. Was it a structural decision? On the other hand, one notices a great freedom in defining the interior spaces. Was this something that came from that decision?

The floors are suspended from the structure, we needed to hang "this" from here [pointing at the plans]. What we wanted was that the building should move a little, like a … here's the gym and here's the street. This is built putting these beams [he draws] and then coming to build here … but this is hanging from there. And this suspension is what enables us to avoid the whole thing being strictly symmetrical.

There are three partial problems: the structure, the façade and the interior spaces, and there's a condition that is this base structure. It's a very difficult programme. And it may have been this simple response to the brief that allowed you to take this step.

Well, let's say it's the two points following those that you mentioned, let's say they are two different moments. That is, the moment of resolving the façade, the infilling of the façade, is not the same moment as resolving the complexity of the spaces. This would lead to something that I've also spoken about on other occasions and which is important. I mean the idea of trying to move freely in confined spaces, the concept of compact architecture. In this case, we return to what I said at the start, how it's misleading to think that this is a simple building, because it's quite the opposite, it's an extremely complex building. But in Columbia this complexity is also energised and activated by the contact the building has to have with the two neighbouring buildings, Pupin and Chandler. And with the street. And with the campus. And all this dictates the kind of multiple attack from possible directions of movement, which the building is nevertheless capable of assuming. And it does so in those spaces that are most lived in. As I was saying earlier, the space for the labs, the teachers' offices, the corridor, and all this creates a unit. And everything is repeated again, with absolute autonomy and freedom. And all this in turn is connected to the stairs, services, etc.

In reality it's also another example of an architecture that shows that it's possible to solve briefs not only by disassembling them, not necessarily by fragmenting its elements, but moving them freely. There is a freedom, perhaps more restricted, but a freedom embedded in a prism or a more simple volume. In that this building is really quite exemplary, easily the most complex I have built. An example of compact architecture is the Houston Museum. But that's only on these horizontal planes. Anyway, in terms of thinking about how architecture is done, the Houston project and this at Columbia are not unrelated, are not far away.

The Materiality of the Project

Do you think there's a methodological complicity amongst the design team, between the architect and the engineers, in this case the structural engineers? Is it necessary for a building to be able, as you say in your latest book, to raise emotions like a sonnet?

I'll start by answering the last part of your question. Let's say that in its closed condition this building has the limitations a sonnet has. Of course, the limits of the sonnet are those imposed by the metre, while here it's us that are imposing the metre. Some of these parallels are never so precise or accurate, but this limitation effectively has something to do with it.

Back to the question of the relationship with the engineer. The engineer is extremely competent, valuable and intelligent, but the complicity, to use your word, arises more out of respect and the sense of ease we felt with him. Complicity doesn't mean coincidence in time, that is to say, we didn't all work together at the same time, but the complicity to which you refer lies rather in trust than temporal coincidence. We didn't think the project through together, we told the engineer what we wanted and he solved it pretty well.

Now I'd like to move on to a topic in relation to the engineers, which I think is important but which we're omitting. I'm referring to the facilities. It's clear that in reality we are finishing the building in regard to construction and architecture. But when I walk through the building, I have the feeling that I've given other people something to do – just as the engineer has given me something to do. I've given other people a space and a structure, a base, rather than a space, which was then filled with so many other things I know nothing about. And there comes a moment when I realise that I have nothing to do with the veins, the nerves, the sensors that this building has. I don't know anything about all the thousands of systems … that is, it's as if our body is an inanimate body, lifeless. The life of the building has been provided by others. In the case of Columbia, this world of facilities goes beyond what we normally find, which are more primary facilities such as electrical, or air conditioning, and are the first breath the building takes, that is not the architect's. But in this case there are many, many things that I don't know, and that's why I like to see the building as a base that takes on its life later, a life that I haven't given it.

There are many machines on the upper floors. How is the composition of spaces done, in technical terms, articulated with the team and the university?

Yes, it was imposed. But what I'm saying is that when I go now, I see the building so full of cables that I can't believe it's so physically full, and I ask myself: how have they managed to leave spaces that can have so many things in? You have no idea. In this regard, it's like when you open the hood of a car. I really don't know who the designer of a car is, if it's the man who makes the body, or who it is… But it's true that this case is different from what happens in a building, but I don't know …, I also realise how, when Louis Kahn or Rayner Banham emphasise so much the fact that on one hand there's the facilities and on the other the spaces, they are suggesting a very different option to that of Columbia, where the facilities are closely linked to the physical reality of the building itself. In that sense, what we are doing is certainly more contemporary than what Louis Kahn would have done when visualising these spaces for the facilities, because the facilities can't be seen when they're occupying the space in such a way.

There were three-dimensional studies to articulate the structure and the façade, all that work can be seen in very great detail.

The structural engineer has been very important. Of course, one thing also happened here, the specific conditions of the building were dictated by the importance given by the university not to lose the gym's tracks. It's an aspect that, if you like, comes into the policy of "correctness". If instead of three basketball courts they had wanted two, the structure would have been much simpler, and the façade would have to have been different. The decision was affected by ideology, if you will. The plot is worth so much, it's so useful, that it doesn't matter about spending a lot of money or make such a complex structure due to its location. It's just that not being able to build vertical columns completely changed the story. The stairwell would have appeared elsewhere, it would have been entirely different. Actually, that was an ideological decision.

Inter-disciplinary Design Methodology

Do you think that this building is a perfect example to illustrate the assertion that the building is an adjustment to the circumstances?

Yes, I think that's very true. That's a quote I found by chance. It seems that Louis Kahn is not the architect who is best represented by that statement. He said that "design is a continuous adjustment to the circumstances". And in some ways, I think that's true. I believe that design is the process of understanding the constraints that are occurring at different levels throughout the design process, that arise in areas that, if you like, are unrelated, but that always leave circumstantial aspects extremely present that are defining of the design. And in this case, the Columbia project is an extreme example of this.

Working with American universities or American institutions has the advantage that you are dealing with clients who know what they want and have the means to spend. And that's a big help for a project. So, building in America may be more trivial, precisely because American society doesn't like to spend money now on construction, plus construction costs are so high that they are really very suspicious of any solution that involves risk and is uncertain.

American construction today is very predictable, for all architects. The architecture of Gehry, for example, is an architecture rich in its formal aspects, but utterly conventional where building systems are concerned. This knowledge that in general American institutions have about what they want makes it easier to do the design, because this is the result of a continuous dialogue. In general, an American institution wants to participate in the project and wants to know and verify that the brief is actually met and that it will satisfy their needs.

Would you say that the institutional client is almost part of the design team?

I wouldn't go that far, but it really helps the architect, the designers, when it comes to the final detail, i.e. in that they are more stringent, they are not so condescending. If they think they may need a space of 600 square feet, they're not convinced when you tell them it can have 200. In this respect, this joint responsibility, because they are adamant that the brief is met, helps a lot. This will also apply to the prices and the cost of the project. Then there's that moment they call "value engineering", which generally allows us to know what the project will be worth. If they see you have a certain amount to spend on the project and recognise that the project will cost more, they look at where they can reduce costs. Thus, projects are born with more construction and less meaning. But there it's almost impossible to make a change during construction. We have to make an effort to get a very precise definition.

Then, the work schedule is as follows: we do the architectural design and then an architecture office makes the working drawings, the details, and there are the many endless "consultings". Maybe 10 or 12, or 15 "consultings" alone for the construction process. There is someone who takes care of the curtain wall, there is someone who deals only with the tightness of the roofs … that is, it involves many specialists. To get everything going is more the concern of the architectural office that makes the "working drawings". So I see no such difficulty, given today's communications and given the separation of the work. But that also means making an effort to listen and to attend all the "consultings". On the other hand, everyone has something to say, and we have to be careful not to let any ideas that you consider key to the project get distorted, with the successive arrival of people.

And in the construction phase, how do you work?

The construction phase is almost … the architect has almost no presence, because practically everything is already defined. It's the opposite of what happens here. On the other hand, American construction is incredibly well coordinated. As I said earlier, in general the constructive substance with which you work is very elementary, American culture these days doesn't value, doesn't have the taste for the utmost quality. This is something that has been lost over the past 40 years. But in 1939 – I've quoted these examples thousands of times – the Rockefeller Center marked a taste for quality of construction, or an academic building like the National Gallery in Washington couldn't be built better or with finer materials. That doesn't happen in America today. Today, even if one wanted to build well, one couldn't because American culture has lost that taste. Surely it's some of the last buildings by Mies or, once again, a building like the National Gallery by I.M. Pei that should be mentioned, but the recent skyscrapers don't have the quality that buildings have had in the past, neither do the buildings of Gehry respond to the technological culture or technological power that is America today.

The construction companies are extremely well-organised. It's miraculous how they're able to meet deadlines and get things done on time. Something you're not used to here. We're accustomed to things not being completed when they should, and this upsets the whole construction process. But there, really, organisational aspects, which are closely linked to the cost of the work and the funding are extremely valued. And so the work almost progresses on its own.

Does that mean there's no room for experimentation with materials such as coloured concrete, etc.?

It's very difficult, very difficult. Well, it's also true that all American projects go through something we do here, which is building a mock-up, a full-scale model. And all this has been tried. The curtain wall has been tested in wind tunnels and everything has been done as a full-scale model before building start, and has been tested in the harshest conditions. So there's little scope for experimentation later.

Do you think such a difference in the relationship to the work between Europe and the US brings something to the project? In Europe and in the US, where the architect almost has to ask permission to go on site, is the experimentation in construction phase lost for the architecture?

These experiments have to be made, as I said, before starting the project. The site isn't the place for experimentation. You can do some experimenting before you start building, but in today's American culture you can't do much experimentation with the construction. In Switzerland, things are not built in the same way as in America, it's very different. But in Switzerland the experience I've had is that of a country where the taste for good construction is highly valued. Switzerland is a country where you can build better, better than in America, I think. Better than in Germany, and better than in Sweden.

Quality is still something that's valued and is thought to have a price and people are not stingy about paying it. The Swiss say it's because everything is so subject to the laws of insurance, that homes and mortgages have to last a hundred years and they can't afford to build badly. That is, the more modest the country the more you can afford things poorly in terms of durability and also in terms of having the architect take a dictatorial role. In fact you see today, I don't know, the significance of architecture has changed in different countries. Architecture doesn't have the same meaning in a highly developed country as in a developing or half-developed country, as we have been in the last twenty years – I mean in Portugal and Spain – or in an emerging country like China. There are significant differences. So architecture is also suffering.

The theoretical disquiet that you show in each project, comes from the time you spent at college or from your training?

We can assume that everyone has certain tendencies, and certainly that in all our careers aspects of our characters have some weight. Maybe in this regard I've been tempted by the humanities and literature since my youth and my adolescence, and surely that is also reflected in my architecture, which is not geared only to the expression of technique. It's an aspect I've never tried to disguise. Neither has Siza, for example. Siza's architecture has never claimed, in an era where it seems that technology is foremost, to make his architecture just that. He has other interests which are there, for everyone to see, in his projects, and also in mine. I think that from his first contact with architecture, an architect can recognise that this desire to make architecture ingrained in the cultural debate of the moment is what he pursues. In this case we have on the table it's hard for me not to "historicise".

And this theoretical foundation, this theoretical rigour, does it later change through constructive discussion during the design process, so that the building responds rigorously to the theoretical problems that you imposed yourself? For example, in this project that you showed us, in which the building was the same size as the existing one and in which the constructive response, when choosing wood, seeks to answer and accept the constraints arising from the first intuition…

I think so, yes, but I don't know if it's always the case. In that respect, surely the techniques and the means available are often ahead of our work. There are times I regret not contributing as much as I would have liked to the development of techniques with which the project will unfold.

But I also think that omnipotence would be dangerous, to think that one can design buildings with a method of forced labour, which would be completely unacceptable, given the social environment and the conditions under which one works. For example, returning to this particular case – and we'll shortly put it aside because it's a project that I started thinking about a fortnight ago – I'm not an expert in timber, and it's something I'll have to learn about when we start to build, to do so efficiently. If I think about this wall here [pointing to the plans], what interests me here is that it is thin in relation to this one. I don't know what the exact measurement for the timber wall is, but what I'm saying here is just, if these walls are 50 and 60 cm because they are walls of masonry and stone, I would like to do these at 20 or 25 cm, i.e. not solid. In a way this would be like a Gothic framework. I would like here to make an allusion to Gothic studwork, and proportions. What we see here in the pictures is still not exactly what I want, but let's say the intention is there.

Do you think, then, that architecture means mastering construction, like you said in a paper?

Not just that. Well, I must have meant to say in some writing how architecture in the past has been a bridge between the arbitrary and the necessary, which means that architecture is responsible for how form appears in the construction. Then, to get back to your question, this mastering of construction occurs only at the moment in which architecture has taken control of it and has been able to give shape to what we build and to match the construction and the form. But it's more the case, I think, of the architecture of the past than of contemporary architecture.

Does this mean a discipline in architecture or architecture as a discipline?

These questions are very long and very tricky to answer, but I think that today we're talking about all of this interdisciplinarity and practising architecture with the input of elements that do not belong. When I speak of architecture with discipline, I mean that architecture is a body which has developed a range of instruments and mechanisms and knowledge which are unique to construction and architecture itself. In this sense, disciplinary knowledge is knowledge that is inherent in architecture and that knowledge seems more important than trying to seek support from the arbitrary help that can be offered by any transformation that converts the interpretation of a physical phenomenon into a building. I think that in the great moments of architecture it has defended itself very well with the expression of what is the body of disciplinary knowledge.

In Iberian construction, there is a lot of experimentation with components, textures, materials … like in the Prado, where the structure behind the cloister has an incredible quality, which I had the opportunity to see a few days ago. How do you view this aspect, is it a need of the architecture, is it a legacy, is this experimentation important, the erosion that has taken place, the colour?

I think some people have said that architecture is the synthesis of all other arts and art shows the need humans have to make visible what we aspire to, what we want to be. In that sense, and going back to what you're wondering, I think that architecture, as the projection of the desires of a particular culture, is very valuable. And in a cultural moment in which construction techniques seem to emphasise only the industrial side, introducing other aspects which offer distinct alternatives and which lead to thinking about colour, for example, interests me. What people like Herzog & De Meuron are basically saying, to mention architects who have both much explored all these aspects of textures and colours in recent years, is that in a world where senses are important, there's no reason to lose them. And this seems to explain much of what has happened in recent architecture, in contrast to all these more aerodynamic architectures, an architecture of form generated purely mechanically, such as some recent projects by Zaha Hadid, or the designs of those trying to extract unknown aspects from other sensory experiences. The Iberian architecture you speak of, Portuguese architecture, places great emphasis on minimalism, on taste in materials and the expression of materials. I mean that here humans also seem to retrieve some contact with this other nature. We really live in a time where, from an ecological point of view, nature seems so enticing, or we would like it to pose so many demands on us, and yet, we see it far away and see what we produce as second nature. Something we want to achieve is that artificiality be lost. Something that demands simplicity of construction.

General Arrangement Drawings

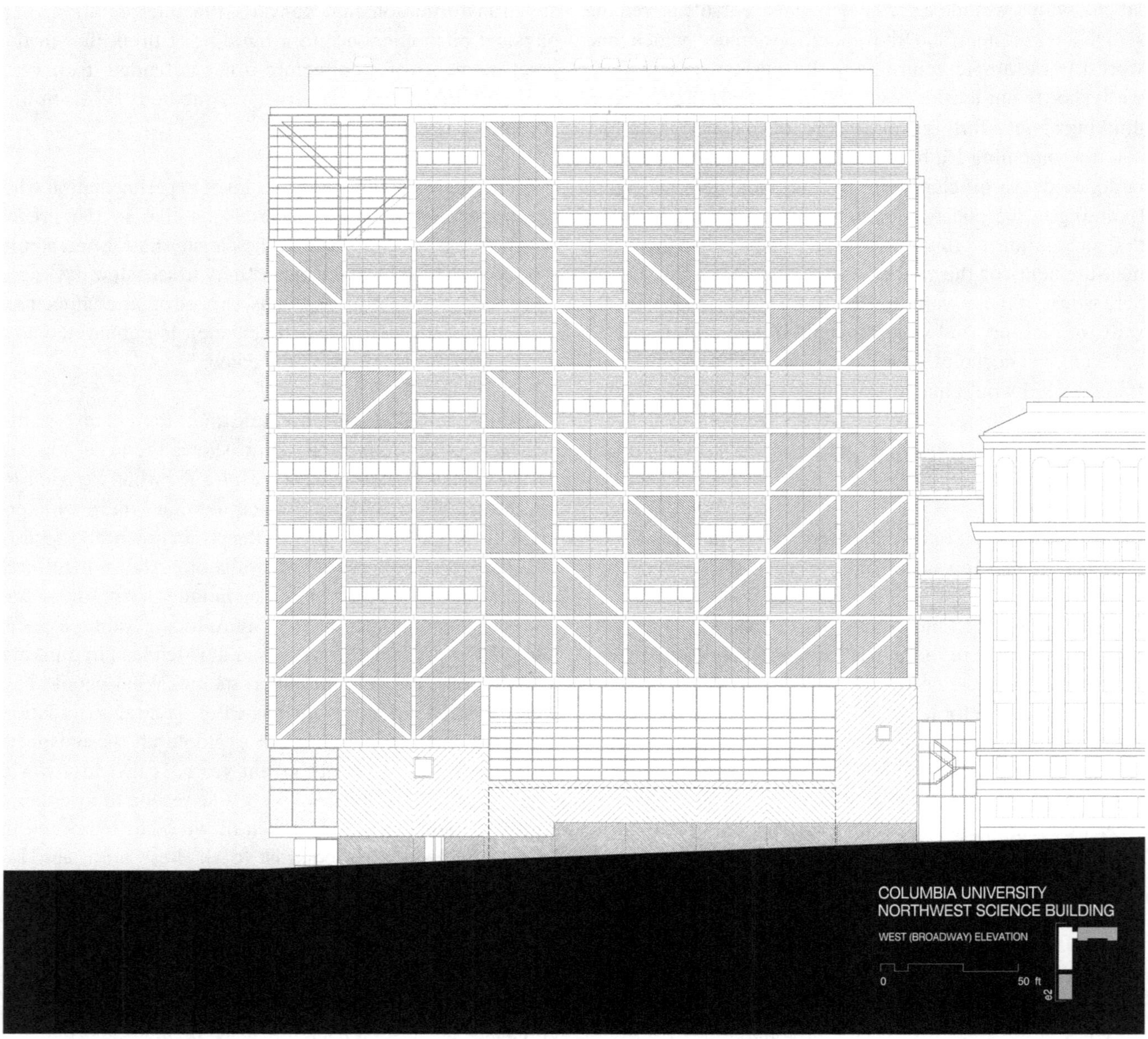

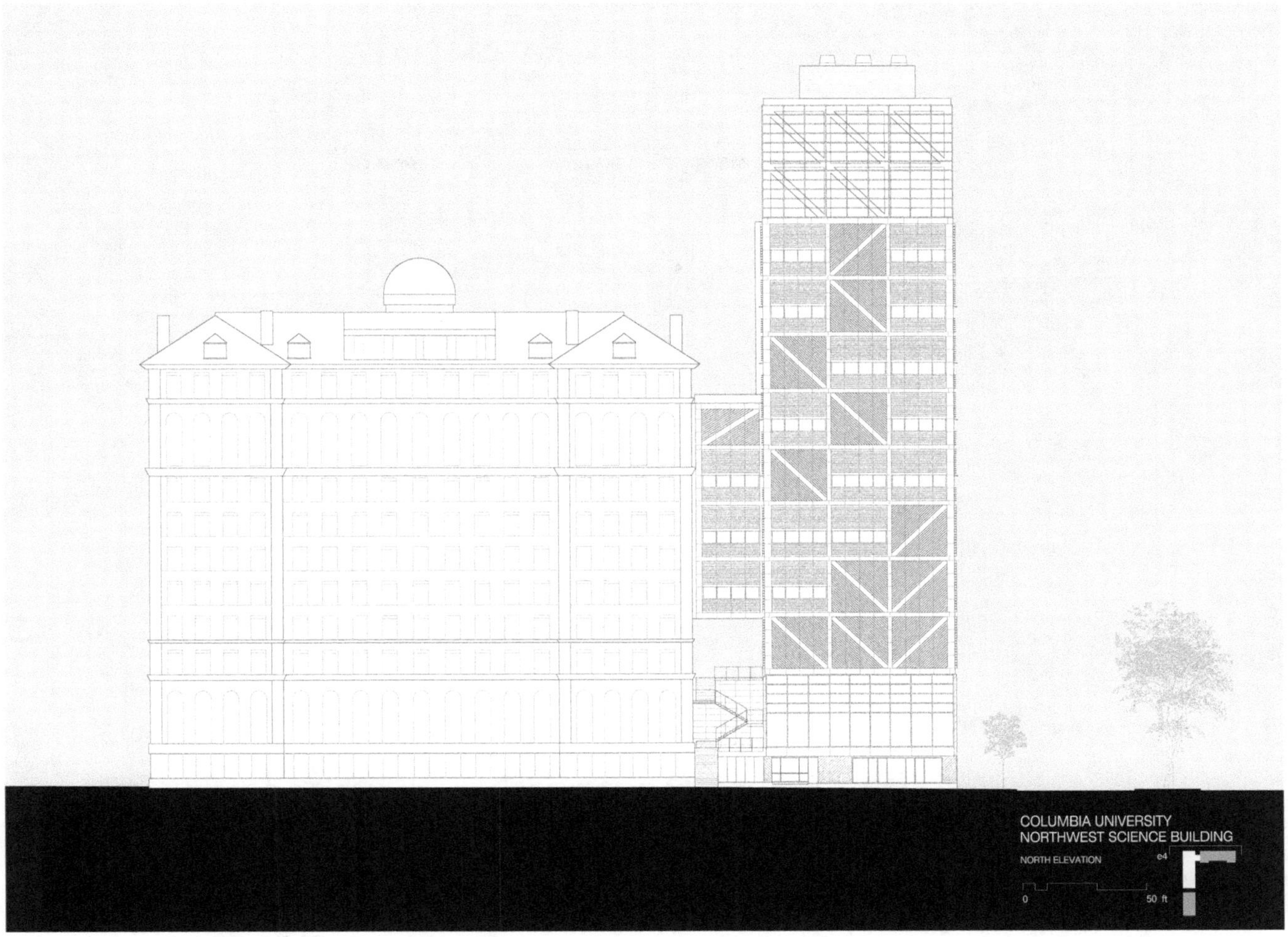
COLUMBIA UNIVERSITY
NORTHWEST SCIENCE BUILDING
NORTH ELEVATION
e4
0
50 ft

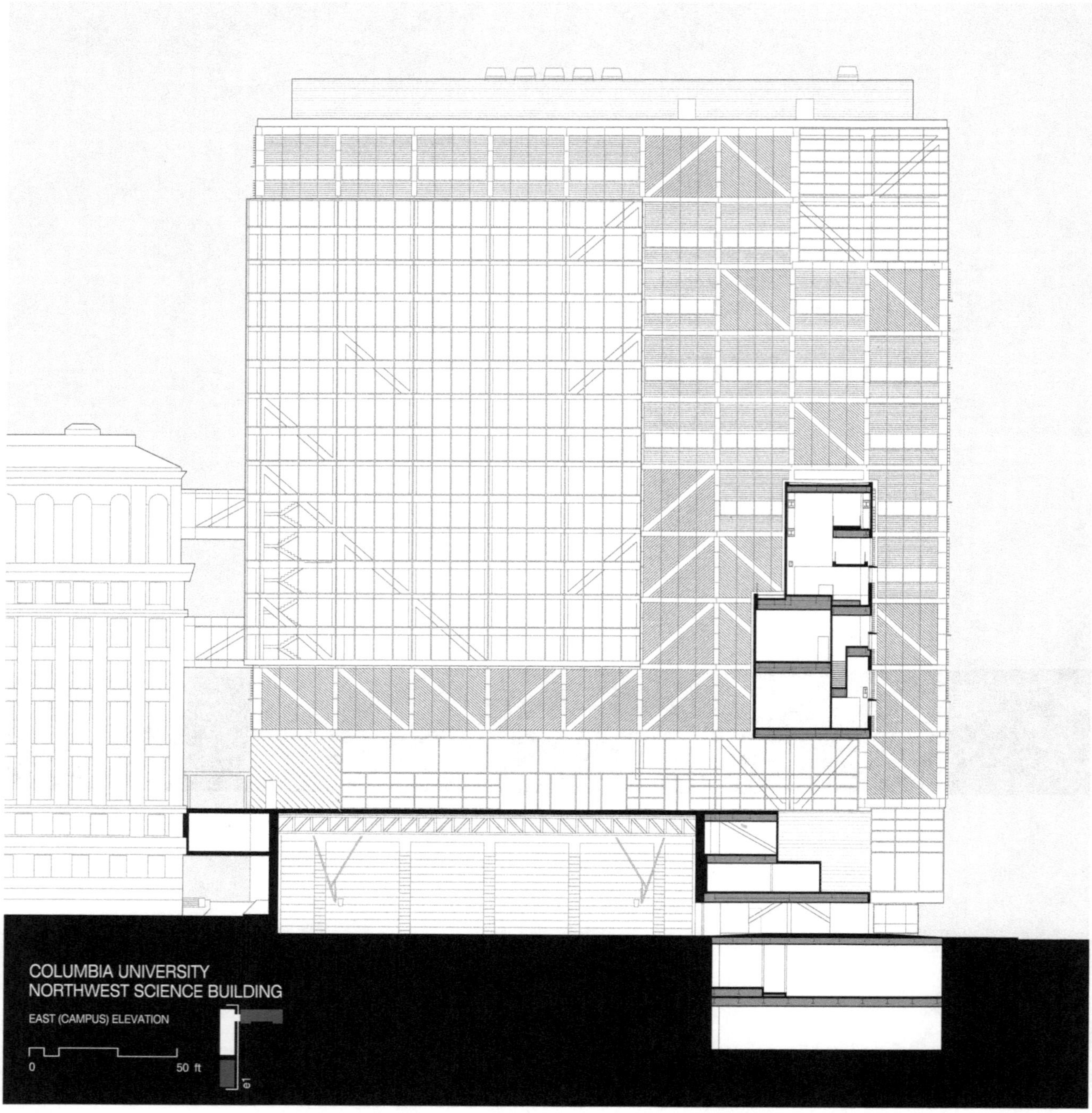
COLUMBIA UNIVERSITY
NORTHWEST SCIENCE BUILDING
EAST (CAMPUS) ELEVATION
0
50 ft

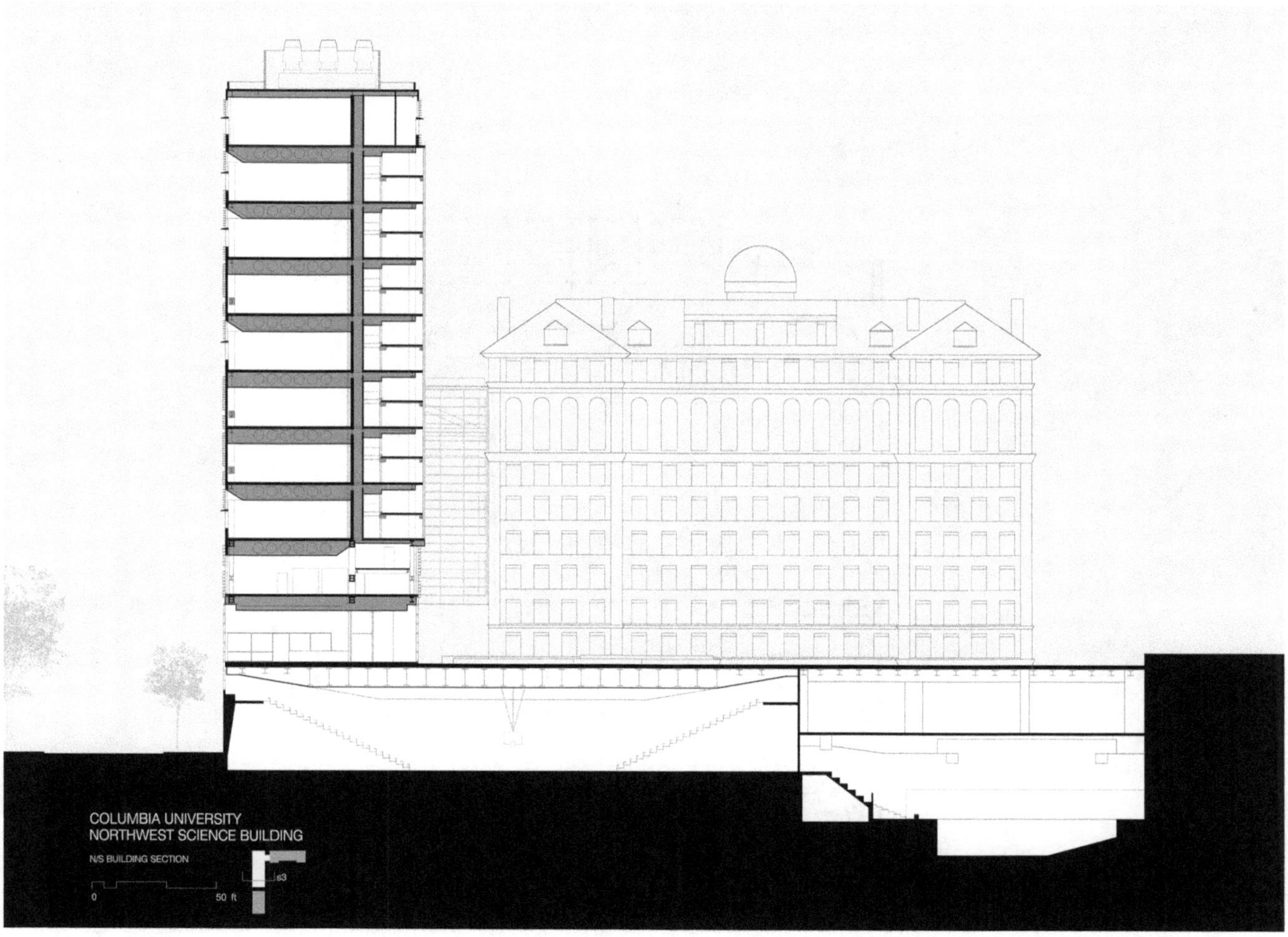
COLUMBIA UNIVERSITY
NORTHWEST SCIENCE BUILDING
N/S BUILDING SECTION
s3
0
50 ft

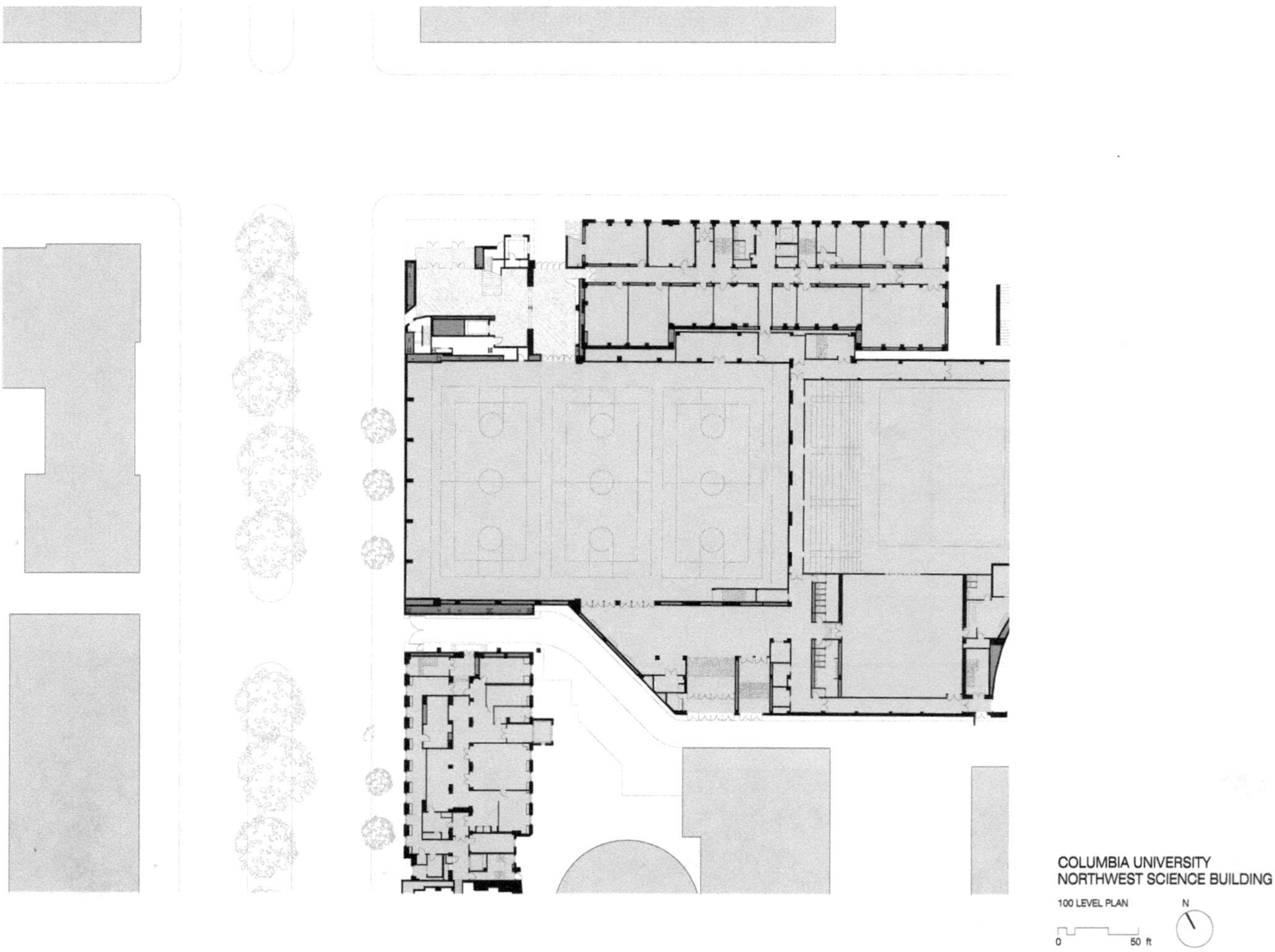
COLUMBIA UNIVERSITY
NORTHWEST SCIENCE BUILDING
100 LEVEL PLAN
N
0
50 ft

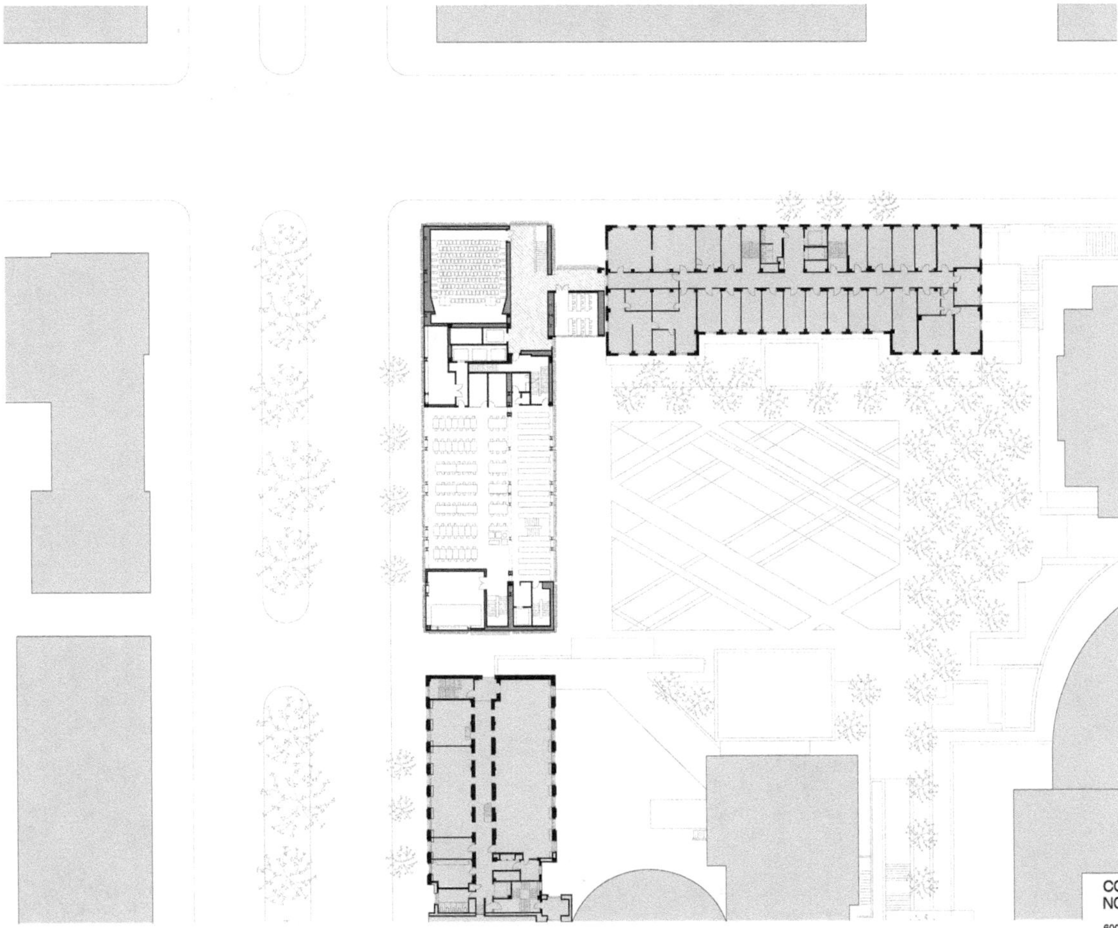

COLUMBIA UNIVERSITY
NORTHWEST SCIENCE BUILDING
600 LEVEL PLAN
0 50 ft
N

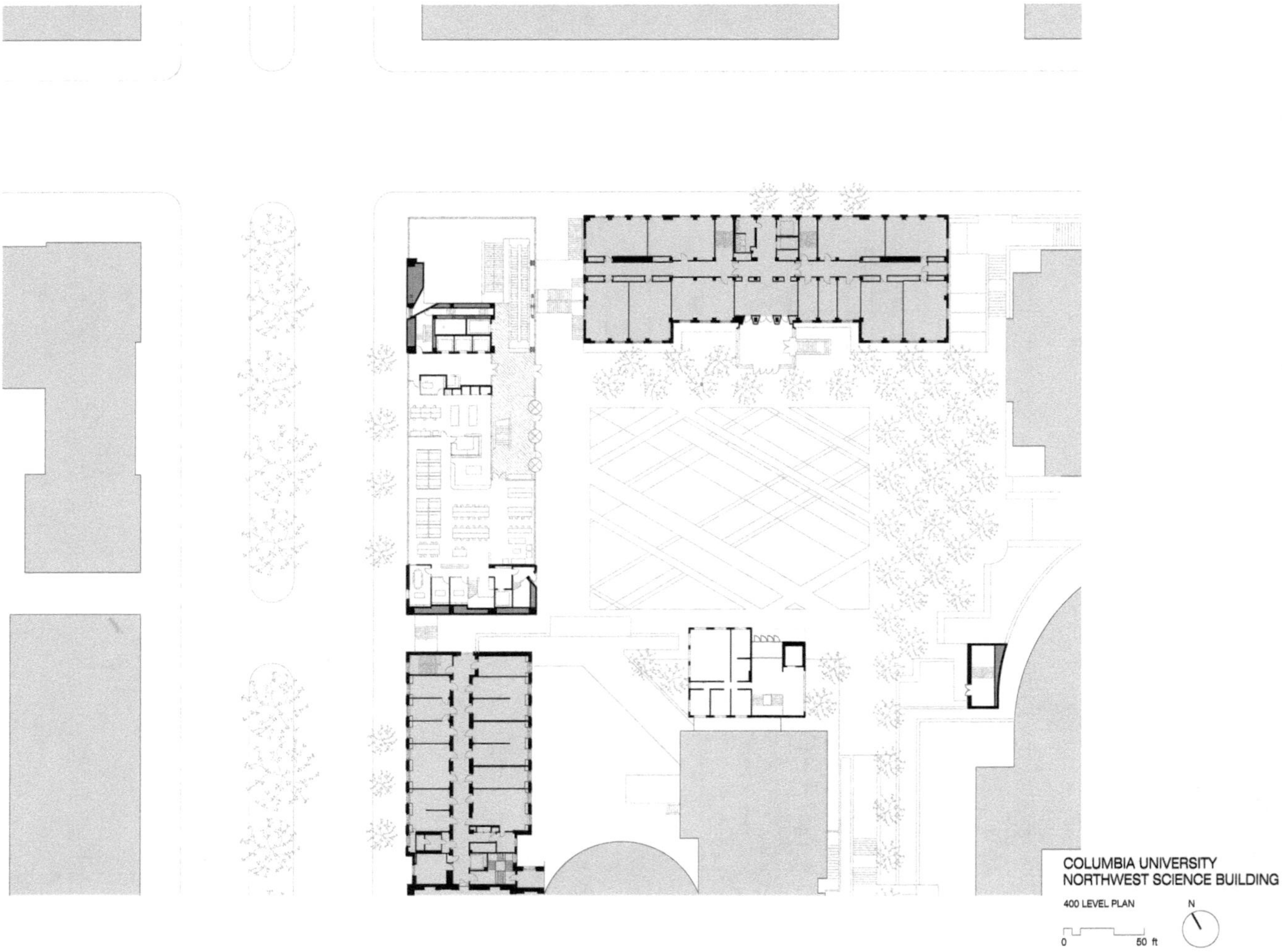
COLUMBIA UNIVERSITY
NORTHWEST SCIENCE BUILDING
400 LEVEL PLAN
N
0
50 ft

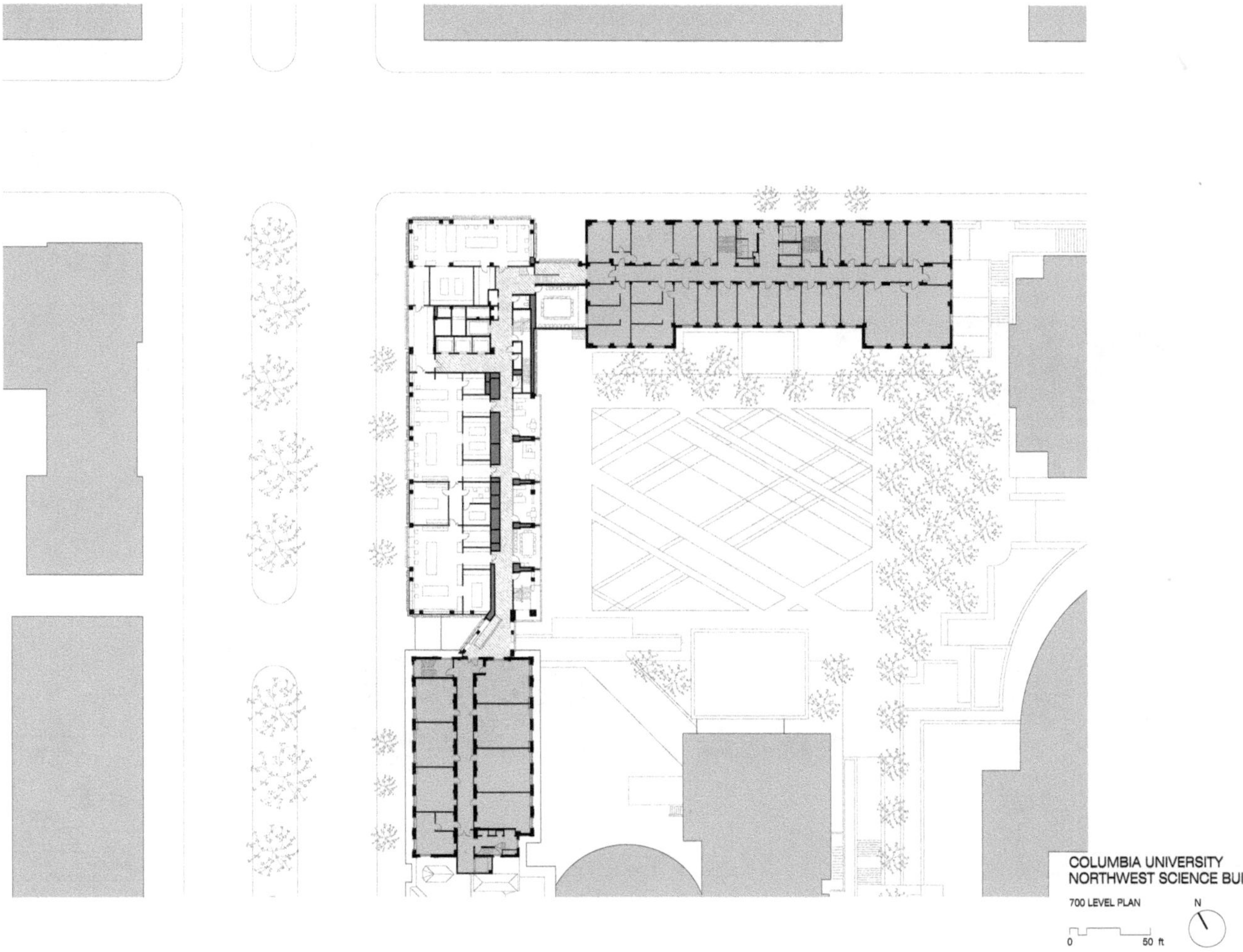

COLUMBIA UNIVERSITY
NORTHWEST SCIENCE BUILDING
700 LEVEL PLAN

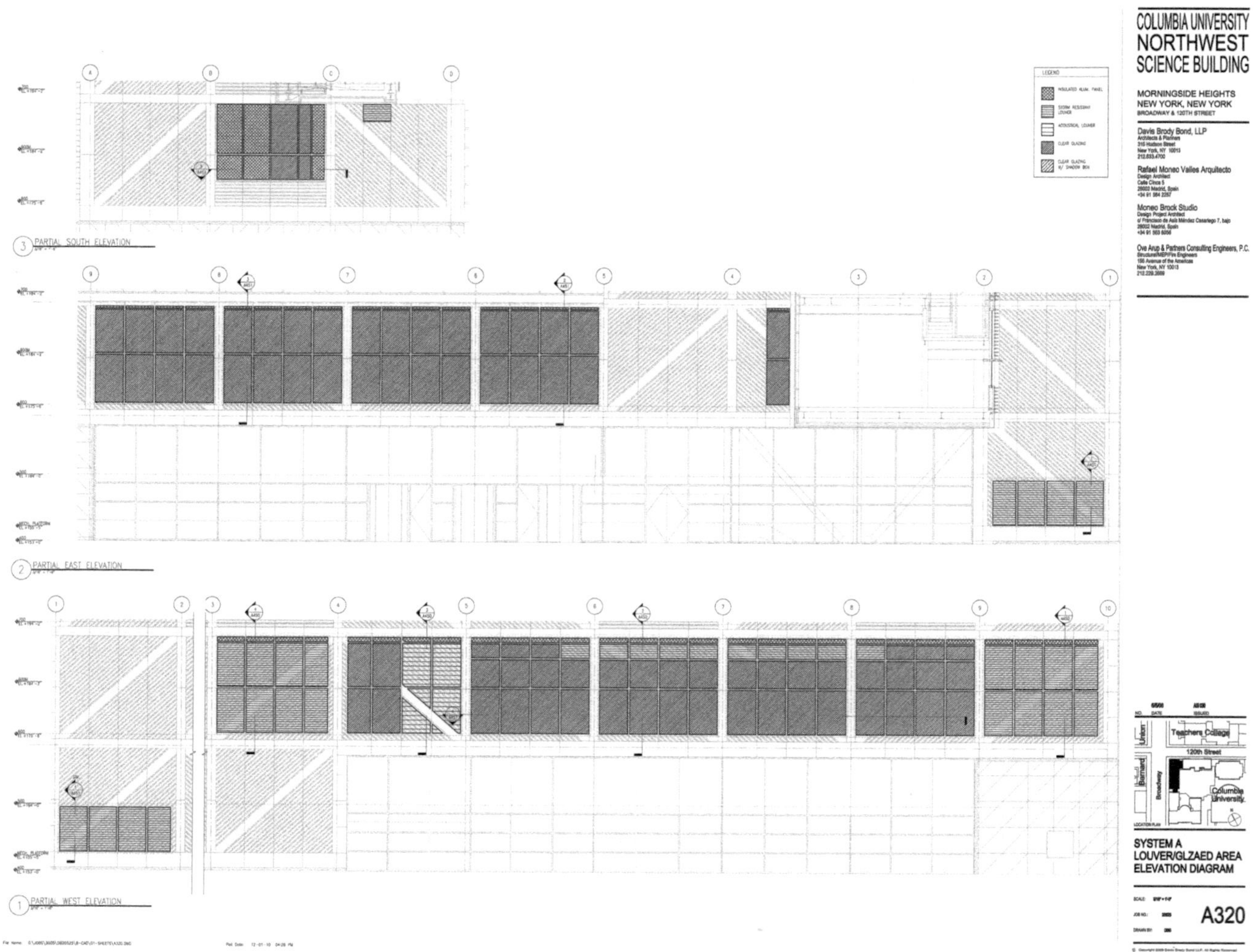
COLUMBIA UNIVERSITY
NORTHWEST
SCIENCE BUILDING
MORNINGSIDE HEIGHTS
NEW YORK, NEW YORK
BROADWAY & 120TH STREET
Davis Brody Bond, LLP
Rafael Moneo Valles Arquitecto
Moneo Brock Studio
Ove Arup & Partners Consulting Engineers, P.C.
LEGEND
3 PARTIAL SOUTH ELEVATION
2 PARTIAL EAST ELEVATION
1 PARTIAL WEST ELEVATION
Teachers College
120th Street
Broadway
Columbia University
SYSTEM A
LOUVER/GLZAED AREA
ELEVATION DIAGRAM
A320

Street View
Copyright Michael Moran

Detail of the street façade
Copyright Michael Moran

View from the courtyard
Copyright Michael Moran

Detail of the glazing in the façade
Copyright Michael Moran

Shading system of the façade
Copyright Michael Moran

Northwest Sciences Building by Rafael Moneo: Circumstantial Evidence

Joan Ockman

Among the suite of images put together by Rafael Moneo's office for purposes of a Powerpoint presentation about his Northwest Sciences Building at Columbia University are a rustic stone-and-brick Basque farmhouse with diagonal timber bracing and a drawing by Jasper Johns from the artist's *Crosshatching* series. Beyond their shared diagonal iconography, these evocative images from utterly different worlds—along with a third, the facade of Mies van der Rohe's unbuilt project of 1953–54 for a long-span convention center—suggest the gamut of inspiration and aspiration that informs the architecture of Moneo's Columbia University building. The brief was a demanding one: to fit into the university's century-old brick-and-mortar McKim Mead & White campus; to house twenty-first century research in several theoretical and applied sciences; to span a large preexisting sports facility that occupies most of the ground and subterranean level (and had to remain open during construction); to bridge to two adjacent science buildings; and to complete the corner of the Morningside Heights campus on a sloping city block while making a mark on the neighborhood skyline and emblematically opening to a planned extension of the university half a dozen blocks northwest in Harlem. These heterogeneous givens demanded a complex, not to say complicated, solution. It is to Moneo's great credit that he succeeded in finding an architectural image coherent and legible enough to subsume them all.

"Finding" is the operative word here. Apropos of Jasper Johns's print, it is interesting to note the origins of the *Crosshatchings* series, which preoccupied the New York artist for almost a decade starting in the early 1970s. Based on a network of hatched lines that change direction and color in a sort of hive pattern, the striated surfaces initially derived, according to Johns, from his recollection of a car passing on the Long Island Expressway: "[A] car came in the opposite direction. It was covered with these marks, but I only saw it for a moment, then it was gone—just a brief glimpse. But I immediately thought that I would use it for my next painting." Another series of etchings contemporaneous with the *Crosshatchings* and often paired with them, known as the *Flagstones*, similarly emerged from an ephemeral but concrete experience. Johns was in a taxi traveling through Harlem on his way to the airport when he passed a store that had a wall painted to resemble flagstones. When he later returned to photograph it, it was no longer to be found, to his dismay, and he was obliged to reconstruct the wall surface from memory. "What I had hoped to do was an exact copy of the wall," Johns recalled. "If I could have traced it I would have felt secure that I had it right. Because what's interesting to me is the fact that it isn't designed, but taken. It's not mine."[1]

Apart from the unexpected appropriateness of Moneo's allusion to this period of work by Johns—whose local inspirations Moneo may or may not have been aware of—the preference for *finding* rather than inventing form is one of the foundation stones, so to speak, of the Spanish architect's approach to architecture. Ever since the beginning of his career Moneo's inclination has been to solve architectural problems by willingly embracing existing circumstances as material and historical necessities. Reality—the "barbarous, brutal, mute, insignificant reality of things," as Ortega y Gasset once put it[2]—is received by the creative individual as a destiny, but one that he or she is able to "reabsorb" biographically and culturally into praxis and, through active "resistance to what is habitual and customary," to utilize in producing "a new kind of gesture," "a new series of realities."[3] This contingent, anti-idealist persuasion, which nonetheless does not preclude a utopian dimension, has been at the heart of Moneo's work for four decades.

J. Ockman (✉)
University of Pennsylvania a School of Design, Columbia University, New York, NY, USA
e-mail: jo2@columbia.edu

[1] Fred Orton, "Present, the Scene of…Selves, the Occasion of…Ruses," in *Foirades/Fizzles: Echo and Allusion in the Art of Jasper Johns* (Los Angeles: Wight Art Gallery, UCLA, 1987), 168–69.

[2] Ortega y Gasset, "The Nature of the Novel," *The Hudson Review*, vol. 10, no. 1 (spring 1957): 30.

[3] Ibid., 33, 36.

V. Abrantes et al. (eds.), *The Pre-Fabrication of Building Facades*, Building Research: Design, Construction and Technologies, DOI 10.1007/978-3-319-22695-8_7

In the case of the new sciences building, Moneo's at once reverent and rational attitude toward reality has resulted in a building that is in every respect a tight fit. While the polished, up-to-date finishes of the interiors appear spiffy and even rich compared to most of the well-worn campus buildings, there was little room for rhetorical excess. The structural challenges of spanning the preexisting (and otherwise unremarkable) gym and dealing with the steep elevational change from the Manhattan street up to the plinth of the campus, together with the highly specific requirements of contemporary lab and research facilities, determined the building's form in both section and plan. Within these constraints, Moneo managed to insert a generous two-story library with warm wood furnishings and good natural light. More suggestive of Scandinavian modernism than high-tech science, it recognizes that students and teachers, whatever the nature of their work, appreciate a comfortable and humane work atmosphere. He also managed to carve out of a strongly raked section an inviting lecture theater and to incorporate a café and penthouse "event space" that take full advantage of their privileged urban views.

On the exterior, the bold and surprising gesture of the diagonalized metallic facade eschews purely spectacular values, nodding more to Mies than to Rem Koolhaas in deploying the hatched grid to express the building's powerful internal structural framework. Framing, to recall Heidegger's concept of *Gestell*, is what the architect and engineer *do* with modern technology when they use it purely instrumentally. Treated merely as means, technology ruthlessly depletes the world even as it constructs it. In conjuring the cultural memory of the vernacular Basque farmhouse in his high-tech facade on upper Broadway, Moneo clearly wishes to insist on a deeper relationship between *techne* and *poeisis* while accepting his building's inevitable participation in the "world picture" of contemporary construction.[4] Beyond its technological presence, the picture plane of the facade also subtly resonates with other kinds of echoes and allusions: besides those already mentioned, to the herringbone pattern of the campus's brick paving stones, for instance, or the louvered, light-modulating facades of Moneo's compatriot José coderch de Sentmenat.

The Columbia building also reflects Moneo's longstanding interest in questions of typology. It is not an accident that it bears a family resemblance to two other buildings that his office has completed in recent years on American university campuses, the Chace Center at Rhode Island School of Design (2008) and the LISE building (Laboratory for Integrated Science and Engineering) at Harvard (2007). Each of these buildings is characterized by the play of an asymmetrical glass-and-metal curtain wall against a more recessive stone or brick ground, suggesting that hybridity itself may well be a typological condition of contemporary academic buildings. Their outer forms reveal the will at once to mesh with and to distinguish themselves from their surroundings and, more generally, to reconcile the disparities of rationalism, urbanity, and scholarly introversion. An iconic predecessor for this building type is James Stirling's likewise hybrid (though considerably more radical) Engineering Building at Leicester University in England (1959–63). Combining Constructivist massing and machinic hyperbole with unexpected contextualism—the diagonally striated glass roof of the lower workshops echoes the pattern made by the rows of housing nearby on the campus perimeter—Stirling's building rears up in its urban fabric as a kind of ship, as Kenneth Frampton and others have noted, embarked on a heterotopian voyage.

The earlier reference to Ortega y Gasset was not incidental. An important early influence on the philosophically inclined Moneo, and a lifelong educational reformer, Ortega located the origin and establishment of scientific knowledge in the agora, gymnasia, and *symposios*—"banquets"—of ancient Greece. He associated this development with Plato's school in Athens: "Plato regards science as a social function and as…a collective creation in which the whole 'city' participates." This wholly social and urban project "requires a special collective organ—which the Romans would call a *socialitas*, or association—charged with promoting it. For this reason, [Plato] founds a *school*." Euclid Later popularized the practice among academics of "living together" and pursuing "investigations in common."[5] Ortega would elaborate on the social and cultural ideals of higher education in his book *Mission of the University* (1930). University education should be dedicated above all "to constituting the type of the whole man," he argued, advancing a prescient critique of disciplinary specialization: "Civilization has had to await the beginning of the twentieth century to see the astounding spectacle of how brutal, how stupid, and yet how aggressive is the man learned in one thing and fundamentally ignorant of all else."[6] While both pure scientific research and professional training remain fundamental to the wider vocation of the university, they should not eclipse its core mission of imparting to students a historically grounded sense of culture

[4] On the relationship between "enframing" and technology, see Heidegger's "The Question Concerning Technology" in Martin Heidegger, *Basic Writings* (New York: Harper & Row, 1977), 287–318. See also Ortega y Gasset's thematically similar contribution to the 1951 Darmstadt Colloquium *Man and Space*, "El mito del hombre allende la técnica," in Ortega y Gasset, *Meditación de la técnica y otros ensayos sobre ciencia y filosofía* (Madrid: Allianza Editorial, 2008), 99–108.

[5] See Ortega y Gasset, *La idea del principio en Leibniz y la evolución de la teoría deductiva*; cited in John T. Graham, *The Social Thought of Ortega y Gasset: A Systematic Synthesis in Postmodernism and Interdisciplinarity* (Columbia, MO: University of Missouri Press, 2001), 417.

[6] Ortega y Gasset, *Mission of the University*, trans. Howard Lee Nostrand (New Brunswick, NJ: Transaction Publishers, 1992), 32–33.

as "the system of vital ideas which each age possesses," "the system of ideas *by* which the age lives."[7]

Thus, on the one hand, the university must not allow itself to be overcome by "the creeping paralysis of scholasticism," according to Ortega; on the other, it must also set aside a place for scientific investigation: "Around the central part of the university, the sciences must pitch their camps—their laboratories and seminars and discussion centers. The sciences are the soil out of which the higher learning grows and from which it draws it sustenance. Accordingly its roots must reach out to the laboratories of every sort and tap them for the nourishment they can provide. All normal university students will come and go between the university and these outlying camps of the sciences." With the outliers of research and professionalism thus buttressing the central institutional purpose of cultural enlightenment, the university should situate itself "[i]n the thick of life's urgencies and its passions… [while asserting] itself as a major 'spiritual power,' higher than the press, standing for serenity in the midst of frenzy, for seriousness and the grasp of intellect in the face of frivolity and unashamed stupidity."[8]

Moneo's sciences building at the northwest corner of Columbia's Morningside Heights campus seems to come very close indeed to Ortega's vision. Perhaps it is not going too far afield here to also remark the convergence between Ortega's "vital reason" and the philosophy of American Pragmatism.[9] From this standpoint we may note that Moneo's building directly faces Columbia Teachers College on 120th Street, an institution historically associated with John Dewey and his ideas of education and aesthetic experience. The diamond-shaped decoration inscribed in the red-brick Gothic Revival facades of Teachers College is yet another of the circumstances surrounding Moneo's "cross-hatching."

[7] Ibid., 60.

[8] Ibid., 77, 81.

[9] For an intellectual biography emphasizing the affinities between Ortega's existential humanism and William James's radical empiricism, see John T. Graham, *A Pragmatist Philosophy of Life in Ortega y Gasset* (Columbia, MO: University of Missouri Press, 1994).

Drawings in the Air

Jeffrey M. Brock

We are very grateful to have been entrusted by Rafael Moneo with the task of drawing up and developing with him ideas for the NWSB. Over the course of 6 years we were at the heart of design development process, making this one of the most fulfilling professional experiences of our careers. Working through issues critical to one's personal philosophy with clients, consultants, engineers, users, city officials and representatives from affected communities, one comes away with the satisfaction of having participated in a truly great architectural production. As dramatic and intense as discussions related to the building design may have at times become, working in such an extensive team, with so many very talented people, was truly invigorating, driving us all to offer our best work. The pivotal role of the architectural teams' leaders, Rafael Moneo on the eastern shore of the Atlantic, Will Paxson on the western, cannot be overlooked.

Having attended the Graduate School of Architecture, Planning and Preservation at Columbia University, Belén and I were more than pleased to become involved in the project.

Of the design process, we would like to convey something of the depth of our investigations into two critical aspects of the building's architecture: the relationship between its overall form and the distribution of program elements; and the development of the building façade. The Columbia building is not the first of Rafael's projects on which we have worked extensively, but the fourth. From the cumulative experience of working on these four jobs, it has become eminently clear that his effectiveness as an architect lies in his knowing perfectly well when a design concept requires vigorous defense, and when one might be better served by demonstrating flexibility and openness, never so much as allowing one's concepts to be overridden, but transformed or evolved. The two questions we will discuss in this article, those of our approach to the building massing and of our treatment of the façade design will demonstrate how this variability of attitude with respect to the negotiation of differences was so effective to the final result.

The first of the major design issues we contended with was, quite naturally, the building massing. For reasons that have been described at length in writings elsewhere, Rafael recognized early that the footprint defined for the building in the northwest corner of the campus in the master plan of McKim Mead and White (MMW) should be respected, despite the municipal zoning regulations allowing for a building with a much greater footprint. The spatial structure of the Columbia campus, the building site's critical position at the perimeter of the campus, and the relation between these conditions and the surrounding city all pointed toward the planimetric conformance with the MMW master plan, and encouraged us *not* to design the building to the limits of the permissible volume envelope as defined by the City of New York. These architectural and urban values were all clear to us once articulated by Rafael, but the pressure we came under to fatten the footprint was intense.

Clearly, real estate in NYC is highly valued, but beyond that, there was some doubt that 65 ft (19.8 m) was an adequate width of floor plate for a laboratory building. The labs themselves required approximately 40 ft (12.2 m.) in their minimum dimension, leaving little room for core and support spaces (offices, workstations, mechanical shafts, circulation, conference rooms, etc.). Our first task was therefore to draw up a plan that worked in order to "sell" the massing concept. Since the labs required 12 ft clear between floors (3.7 m), Rafael suggested the introduction of a mezzanine in the support areas, effectively doubling the area available for offices and open work areas. The extra height required to fit a mezzanine on each lab floor was certainly welcome in the single-height areas of open laboratories, but the effect on the overall height of the building was, by contrast, an issue, as we were determined to keep the total height of the building reasonably close to that of its nearby neighbors at the north end of the campus. In order for the Mezzanine concept to be

J.M. Brock (✉)
Moneo Brock Studio, Madrid, Spain
e-mail: jeff@moneobrock.com

V. Abrantes et al. (eds.), *The Pre-Fabrication of Building Facades*, Building Research: Design, Construction and Technologies, DOI 10.1007/978-3-319-22695-8_8

carried, therefore, the clear height in each mezzanine had to be reduced to bare minimums, putting great pressure on the engineers responsible for structure and mechanical systems.

Once the lab plan was shown to work well within the limits of the footprint, we concentrated our efforts on the public spaces occupying the first six floors of the building. Here the architectural problems were laden with import. On its lower levels the building was to serve as a gateway to the campus and a beacon to the community, at once demonstrating the University's openness while at the same time providing an exemplary public space and public amenity. The problem of making all the program elements and building systems fit within the space available was at its greatest at the "head" of the building, where the campus entry was to be fashioned at the corner of Broadway and 120th Street, but where the building floor plate reduces to a square 65' (20 m) on a side as the building volumes slips down alongside the gymnasium. In this small square we needed to pack massive banks of mechanical and electrical systems risers coming from the basements below and serving all upper floors, vertical circulation, as well as the major elements of the building structure (carrying more than half of the entire building dead load, all of the north-south lateral loads, and half the east-west lateral loads), while at the same time creating entry and café spaces that could be characterized as open, airy, and inviting.

To this we added a demand of our own, springing from the University expressed wish to make the northwest corner site an inviting gate: that the street at the corner and the campus above be, to the greatest extent possible, visually connected across all these spaces. We decided that the sense of openness that the University sought needed to be realized in what was to be built by our carving a view corridor diagonally up, penetrating the building volume, opening the slabs and walls in just such a way that the continuity of space could be perceived from either end of this passage, so that the route up and through this part of the building would be self-evident and inviting.

In the interstitial space between the new building and its eastern neighbor, Pupin Hall, we were able to introduce an outdoor stair, echoing the indoor path and thereby reinforcing it with a parallel movement out of doors. In the end it is this cascading flow of open spaces across four levels, covering 20 m vertically and 40 m horizontally, that makes the public part of the building come alive. Indeed, it was in the development of the design of the public areas that the wisdom of Rafael's first impulse, to maintain the MMW footprint, was demonstrated. Although the packing-in of the building program into the reduced floor area presented an architectural challenge, the fluidity of movement between levels and the openness of these spaces to the campus and the street all along the path between the ground floor and the sixth would not have been possible without the wide margin of open space left between the new building and Pupin. The decision to maintain the narrow footprint and interstitial space between building as drawn by MMW has been shown worthy of all of the vigorous defense that it required.

Of the building's structural system and its expression on the façade, much has been said, but perhaps we could here emphasize the architectural qualities we perceive, and give some accounting of the development of the façade system. In many ways the building's expression is indebted to the structural system, as the latter embodies what could be considered high ideals: it is both discreet and efficient. Interestingly, while it communicates a sense of flow and strength, its legibility as a refined, optimized structure is mixed with a curiously abstract quality that stands quite apart from its functionality. While the structural engineer's algorithms created the basis of the "drawing" of the façade, distributing the frame's diagonal braces according to predefined values, it needs be said that there was a certain amount of tweaking of the drawing for architectural effects, in some cases for an improved distribution of windows, since it had been decided that these would be located only in bays lacking diagonals, and in other cases for none other than the aesthetic impact the location and orientation of the diagonals relative to one another effect.

In this sense the structural design, optimized first in its use of steel to conserve material and cost of assembly, and later in its architectural aspect, was from the outset recognized as the "defining gesture" of the project. The scale of the building's span over the gym, together with the exacting requirements for structural stiffness engendered by the laboratory function, required a special solution. Having started with a concept of a massive truss being erected over the gym roof to support a standard building frame above, the evolution of the structural design to the fully-braced frame represented the fruit of a highly concentrated collaboration between architects and engineers, bearing evidence to the centrality of the structural solution to the building's architecture. That the final arrangement incorporated a subtle, even subdued efficiency convinced us all that the elegance we sought had been found. That the structural design process was as collaborative as it was demonstrates Rafael's flexibility and openness when a design problem clearly admits contributions from talented architects and engineers.

Clearly, while the urge to reveal the braced-frame solution on the façade was a natural outgrowth of our enthusiasm for it, it was also clear that the steel frame itself would in general not be visible on the facade, but rather cloaked behind fireproofing protection. Thus no sooner had the frame been defined than the question of the building's cladding became critical. Rafael's impulse was to translate the frame's dimensions into the very units of construction, thinking of the voids between the frame elements, a series of identically sized squares and triangles, as the essential space of construction, so making manifest the primacy of the frame. For this reason he has spoken of his wish to apply a technique of infill, calling up images of half-Tudor construction methods, where masonry panels are erected in the spaces between elements of a prominent heavy-timber wood frame, recognizing this centuries-old method as apropos in our own situation.

After studying the alternative means by which the façade might be completed, including stick-framing between floors (infilling), masonry infilling, and precast concrete panelization, the practical realities surrounding the cladding of a tall building in the XXI Century left us little option but to panelize the façade and design it as a curtain wall. The expression of the structural frame design on the façade would therefore have to be a representation of a kind, given the typically rectilinear nature of the panel modules, spanning floor-to-floor and being just 4–6 ft wide (1.2–1.8 m), where our structural bay was 21.5 ft (6.6 m). Our first clue as to how the redrawing of the structural frame geometry on the façade panels might be made came precisely from Rafael's earlier concept of infill: the material quality of the panels could be given emphasis in the spaces between the frame members as they appear redrawn on the skin, through the application of color, texture, or some notable quality of physical presence. In a nod to the technological fruits of scientific research, we opted for a material we thought "sympathetic" to the building program (high-tech science research): an anodized aluminum skin, with a large-scale texture made of applied extruded aluminum fins to give a ribbed or deeply corrugated effect.

In adding aluminum fins in a dense pattern across the vast majority of the façade area we found ourselves engaged in battle to assure the detail had sufficient depth to create the effect we sought. The façade "drawing" was to be made for all intents and purposes exclusively with texture, and the depth of that texture and the shadows thereby created therefore needed to suit the enormous scale of the operation. Curtain walls are, in general, rather flat affairs, with little depth to speak of, and we knew that in developing our system, we were asking the manufacturers and installers of the wall to consider doing things that were out of the ordinary. But with many specialists involved, from façade systems designers, quality architectural metalworkers and curtain-wall fabricators, all coordinating their efforts and offering their best ideas, the end result came together quite flawlessly.

The drawn structural frame on the façade is idealized. Where each member within the true structural frame is sized to carry only the loads it must, in redrawing the frame on the façade we set every member's projected thickness at a uniform 18″ (46 cm), except at the edges of façade-wide panels (the building corners and the top, bottom and side edges of the aluminum areas of façade) where we widened the virtual columns and beams to 2′3″ (69 cm). Interestingly, the redrawing of the frame on the façade panels is a "negative" drawing, in the sense that it is made by leaving *off* the texture of fins in the areas covering columns, beams and diagonal braces. The fin texture is applied to the (virtual) infill panels, acknowledging the importance of the structure by stopping at its edge, leaving a void as a mark, indicating where the frame lies behind. The presence of the frame, which is itself not visible but hidden behind the cladding and the fireproofing further beneath, is made evident by the absence of fins, a presence made evident through absence, a kind of drawing through erasure.

On the panels covering structural modules containing braces, the fins are turned to run in parallel to those diagonal braces, while over the brace-less modules the fins run horizontally. This tripling of the fin orientation creates a patchwork of different light effects on the façade, effectively reducing its apparent expanse. The three-dimensional form of the fins and the layout and dimensioning of the jointing between the panels was studied in depth with the metalworkers and curtain-wall manufacturers to give the desired architectural effect, while staying within reasonable working methods. The fin solution proved highly adaptable, allowing for the location of mechanical systems grills, diffusing light sources and vision windows for the interior spaces.

In the end, working out the form of the building and the details of its material composition, given the conditions we faced in New York (with tight constraints of space and cost, besides the hefty structural challenges) made working on the Northwest Sciences Building an exhilarating experience. By clearly seeing and identifying which design imperatives were most important to defend, Rafael inspired great work from architects and engineers from all sides. By developing a forceful design concept of space, movement, and relationships at the scale of massing and space planning, and an equally forceful concept of the expression of the building's own architecture as carrier of meaning in the façade design, he led a terrific team to produce a building that is elegant, brave, technical, modern, clean, and rational, among many other things.

Acoustics and Noise Control Within School Buildings

Diogo Mateus, Andreia Pereira, and Vítor Abrantes

Introduction

The last few decades have experienced robust developments in technology and an increasing demand for physical comfort in all sorts of settings. In particular, acoustics and noise control issues have assumed an important role in buildings of all types. Our focus here is on school buildings in particular. In Portuguese school buildings, the use of heating, ventilating and air conditioning systems (HVAC) has been intensified, due to requirements mandated by Portuguese regulations of energy systems and air-conditioning in large buildings (RSECE). As a consequence, the increased production of noise provided by HVAC equipment has been affecting not only the learning process during classes but also the neighbouring population, causing acoustic annoyance. In many cases, the network of HVAC ducts creates breaks among interior spaces of the buildings, causing a decrease in insulation of partitions.

Within this context the present paper discusses, on the one hand, some issues related with the construction and rehabilitation of school buildings with respect to acoustic and noise control problems and, on the other hand, some difficulties in fulfilment of the acoustic requirements established in the Portuguese acoustic code (RRAE) [1], including typical problems found during construction which strongly influence the final results.

D. Mateus (✉) • A. Pereira
Faculty of Engineer, Coimbra University, Coimbra, Portugal
e-mail: diogo@dec.uc.pt; dm@contraruido.com

V. Abrantes
Faculty of Engineer, Oporto University, Porto, Portugal
e-mail: abrantes@fe.up.pt

Behaviour of Sound in Rooms – Main Topics

In general, to provide acoustic protection of buildings and particularly for the case of school buildings [2], as a way of giving an adequate guarantee of acoustic comfort in its interior, it is necessary to act in four ways: provide airborne sound insulation of partitions, either between interior spaces or between the exterior and interior of buildings; provide impact sound insulation of noise produced inside buildings; analyse the behaviour of sound inside lecture rooms, also called room acoustics; noise control provided by mechanical equipment.

Airborne sound insulation between rooms depends not only on the separation element which provides direct sound transmission but also on the contiguous partitions which produce flanking sound transmission. The increase in airborne insulation may be attained by increasing the density of the partition and/or by using partitions built of several layers without rigid connections between each other.

Transmission of impact sound depends on direct sound transmission (when adjacent rooms of different floors are assumed and the noise is generated in the room above), but also on flanking sound transmission by contiguous partitions.

To reduce the impact sound transmission it may be possible to use resilient coverings or floating systems applied over resilient underlayers.

The analysis of sound inside a room depends on the geometry of the room, lining materials, furniture and occupancy. This analysis aims to provide a proper acoustic environment taking into account the volume of the space and the use of the room.

As for noise control of the mechanical equipment of a building, the chosen actions depend on the three mentioned ways of reducing/controlling sound. In particular, for the case of HVAC systems, it is usually necessary to reduce structural vibration transmission and airborne sound transmission through ducts or through the exterior medium.

V. Abrantes et al. (eds.), *The Pre-Fabrication of Building Facades*, Building Research: Design, Construction and Technologies, DOI 10.1007/978-3-319-22695-8_9

To reduce vibration it may be necessary to use resilient elements or floating systems similar to those solutions used in the reduction of impact sound transmission. Here, however, low frequencies play an important role and therefore thicker resilient elements need to be used (or even anti-vibration mounts), of about 10 times greater thickness than those used to provide impact sound insulation. To control airborne transmitted noise through ducts, attenuators may be used or acoustic barriers enclosing the equipment, made of phono-absorbent material.

Portuguese Acoustic Requirements

The acoustic demands that buildings in Portugal should satisfy are defined in General Regulation on Noise (RGR approved by Decree Law n° 9/2007 of 17 January) [3] and by the Regulation on Acoustic Requirements for Buildings (RRAE), approved by the Decree Law n.° 129/2002 of 11 May, with new wording given in Decree Law n° 96/2008 of 9th June [1]. The RGR defines, in a general way, the politics for prevention and noise reduction in order to guarantee a healthy environment for the welfare of the population. The RRAE defines a variety of acoustic demands for buildings in order to have better acoustic conditions inside the different spaces.

The RRAE applies to several types of buildings, defining in art° 7 a set of acoustic demands for school and research buildings or buildings with similar functions (see Table 1) to be built or rehabilitated. For these buildings, this regulation does not differentiate among new buildings and existing buildings that are subjected to rehabilitation works.

In the case of auditoriums, conference rooms and polyvalent rooms, with activities where speech is important, one should consider the demands defined in art. 10-A, which is applied for auditoriums and halls. This article states limits for the average reverberation time T, which are significantly lower (more restrictive) than those indicated in art. 7° for lecture rooms in school buildings (in general the average reverberation time is situated between 50 to 80 % in relation to the limits for a classroom, assuming typical dimensions of polyvalent rooms and auditoriums in school buildings). To accomplish this limit a significant amount of sound absorption must be added to rooms, by covering walls and ceilings with sound absorbent solutions, which in many cases may increase the cost of construction works and may also require the installation of electro-acoustic systems to ensure sound amplification which is not practical to use during lecturing activities and could be therefore avoided.

In accordance with art.° 21 of the General Regulation on Noise and for the case of schools, noise sources present in these buildings that may cause annoyance are subjected to exterior exposure limits indicated in art. 11, as well as in alinea 1b) of article 13 (to avoid exterior acoustic annoyance).

Examples of Constructive Solutions Used in School Buildings

Acoustic protection of school buildings is attained by acting within the four ways defined in section "Behaviour of sound in rooms – main topics", by an adequate choice of the materials used and their correct application.

With respect to airborne sound insulation, one should distinguish between airborne sound insulation between the exterior and interior rooms, where facade sound insulation is usually analysed, and sound insulation between rooms of the building. For the majority of schools, either in new lecture classrooms or in rehabilitated ones, façade sound insulation depends essentially on the window solution, mainly on the type of frame and glass (insulation may also be dependent on the blind box used). Airborne sound insulation between rooms depends on the separation element, on the flanking elements and also on other constructive elements such as ducts crossing the walls or electrical boxes embedded on the separation partition.

Fulfilment of the acoustic requirement for facades D2m,nT,w + (C or Ctr) ≥ 33 dB, for classrooms, teachers rooms, administrative rooms, polyvalent rooms, medical offices and libraries is easily attained with common solutions of windows (for example double glass with thicknesses 8 + 6 mm in a frame with an air permeability class A3). However, in some situations, such as when the percentage of windows area is great and the rooms are of small or medium size, it may be necessary to use better performance solutions to fulfil the acoustic requirement.

As for airborne sound insulation between spaces, in the majority of school buildings, where heavy elements (heavy or lightweight concrete slabs and brick or concrete masonry walls) are used, it is not difficult to fulfil the acoustic requirements defined in Table 1. Brick masonry walls, 20 cm thick, allow accomplishment of the acoustic demand DnT,w ≥ 45 dB, while double walls with 11 + 11 cm, with an air gap filled with a sound absorbent material, allow accomplishment of the requirement DnT,w ≥ 50 dB. Note that for masonry walls, more important than the thickness or type of bricks, it is important to assure that bed and head joints are conveniently filled with mortar, with particular attention to the last joint which connects with the top slab and to ensure that the mortar lining of the wall has a proper thickness (not less than 1.5 cm).

Table 1 Acoustic demands for school and research buildings (Art. 7° of RRAE) [1]

Ref.	Element/space	Acoustic demand		
1a)	Between the exterior and receiving rooms	D2m,nT,w + (C; Ctr) ≥ 28 dB, in quiet zones defined in the alínea b) of the n.°1 of art. 11 of RGR		
		D2m,nT,w + (C; Ctr) ≥ 33 dB, in mixed zones or quiet zones defined in alineas c), d) and e) of the n.°1 of art. 11 of RGR (or to non-classified zones).		
		C or Ctr, added to D2m,nT,w, when the window area represents more than 60 % of the facade (depending on the dominant noise in the emitting space).		
1c)	In receiving rooms[a], noise source from other places of the building	L'nT,w ≤ 60 dB if the emitting room is a corridor with a lot of movement, gymnasium, canteen or technical classroom		
		L'nT,w ≤ 65 dB if the emitting room is a lecture, nursery or polyvalent room		
1d)	Average reverberation time (500, 1000 and 2000 Hz), T, with furniture and no movement	$T \leq 0.15 \times V^{1/3}$ [s] in lecture rooms, polyvalent rooms, libraries, canteens and gymnasiums		
1e)	Average equivalent absorption (500, 1000 and 2000 Hz), A, provided by linings of atriums, corridors with great circulation	$A \geq 0.25 \times S_{plan}$		
1f)	In receiving rooms the value of LAr,nT provided by the noise generated by equipments of the buildings should be:	Libraries		
		LAr,nT ≤ 35 dB(A) if the operation is not continuous		
		LAr,nT ≤ 30 dB(A) if the operation is continuous		
		Other receiving rooms[a]		
		LAr,nT ≤ 40 dB(A) if the operation is not continuous		
		LAr,nT ≤ 35 dB(A) if the operation is continuous		
1b)	Receiving room / Emitting room	Lecture room (including musical), teachers, administrative	Libraries and medical offices	Polyvalent rooms and nurseries
1b)	Lecture rooms, teachers and administrative	DnT,w ≥ 45 dB	DnT,w ≥ 45 dB	DnT,w ≥ 45 dB
1b)	Rooms for music classes, polyvalent rooms, canteen, gymnasiums and technical classroom	DnT,w ≥ 55 dB	DnT,w ≥ 58 dB	DnT,w ≥ 50 dB
1b)	Nurseries	DnT,w ≥ 53 dB	DnT,w ≥ 55 dB	DnT,w ≥ 48 dB
1b)	Corridors with great circulation	DnT,w ≥ 30 dB	DnT,w ≥ 35 dB	DnT,w ≥ 30 dB
		+15 dB if a door doesn't exist	+15 dB if a door doesn't exist	+15 dB if a door doesn't exist
4 and 5	In situ measurements whose aim is to verify the acoustic demand one should consider:	+3 dB for D2m,nT,w and for DnT,w		
		– 3 dB/dB(A) for L'nT,w and LAr,nT		
		– 25 % for T		

[a]Receiving rooms – Lecture rooms, Teachers rooms, administrative rooms, polyvalent rooms, nurseries, medical office, libraries

In the separation between floors, it is also easy to accomplish the acoustic demand DnT,w ≥ 45 dB, but, in the case of lightweight slabs it is hard to fulfil the requirement DnT,w of 50 dB, without an acoustic reinforcement, either in the floor or in the ceiling. For situations with the acoustic demand consisting of DnT,w greater than 50 dB, where the flanking sound transmission strongly influences behaviour, it may be difficult to implement adequate constructive solutions, especially in cases of non-intrusive rehabilitation.

Figure 1 displays, as an example, a layout of the principle of a proposed acoustic constructive solution for separation between a common classroom and an adjoining music classroom. This solution, apart from the concerns regarding sound insulation, comprises also an absorbent system for ceilings and walls of the music classroom.

To isolate the impact noise generated in the floors of school buildings, the solution is usually either a floating screen (see floor system in Fig. 1), over which the floor covering is applied or by using a flexible coat, namely a vinyl or linoleum with a flexible basis. The first type of solution, although compatible with most floor finishings, is not always feasible when remodelling buildings, since it implies a rise in

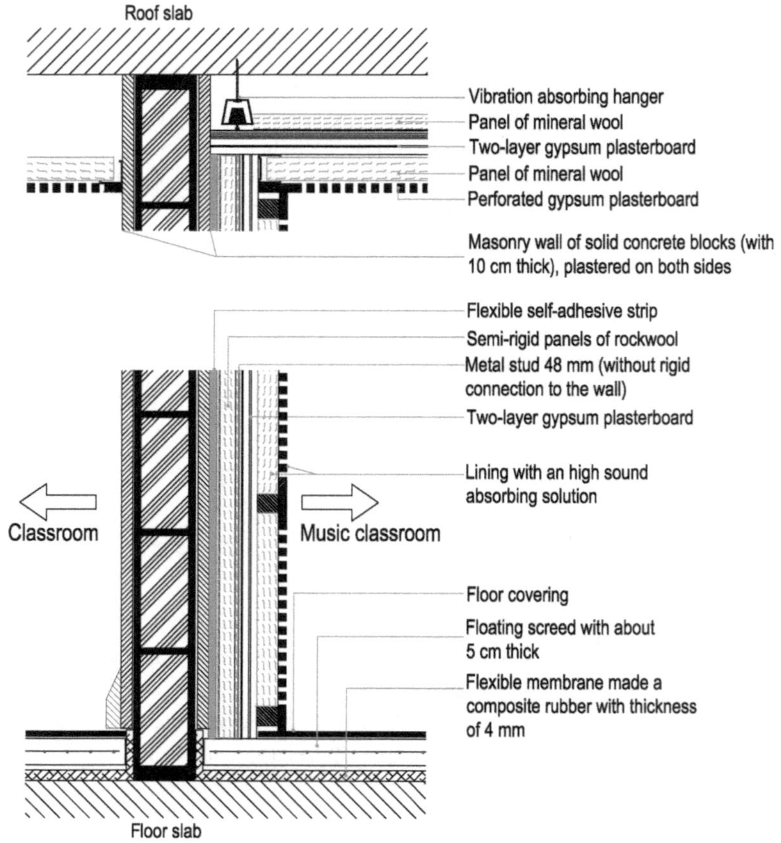

Fig. 1 Example of a constructive solution proposed for the elements that separate a classroom from an adjoining music classroom

the thickness of floors, with a consequent reduction in the height of the ceiling, which is already in some schools significantly lower than desirable.

The second type of solution, which can be applied to existing floors, occupying only a thickness of about 3 to 4 mm, is a bit limiting and usually not viable outside classrooms and libraries, for instance in atriums and corridors of wide movement.

Both solutions, when properly applied, can lead to results that satisfy regulation requirements. However, when using the solution of floating screed it is common to make small execution mistakes, including the creation of rigid connections between the floating slab and the wall, in the plinth area, which strongly compromises the effectiveness of the solution (see Sect. "Typical execution problems"). The option of not applying these solutions in corridors of great movement usually leads to non-compliance with acoustic requirements, but they may be acceptable for cases of non-intrusive rehabilitation, since the beginning and end of classes are generally coincident in several classrooms, of the same part of a school.

The control of reverberation in a school is achieved by imposing an upper limit for the average RT for the majority of spaces or a lower limit for the sound absorption equivalent area provided by the linning materials of corridors with wide movement. Moreover, this solution provides not only a better sound quality inside these spaces but also allows minimize the noise produced and transmitted. For most spaces of a school, although not always corresponding to the optimal solution, the control of reverberation can be done using only the application of false ceilings or ceiling absorbent coverings, e.g., using perforated gypsum plasterboard panels, with air gap partially filled with mineral wool (as the false ceiling of the classroom, in Fig. 1).

The application of these solutions is usually feasible, even in non-intrusive rehabilitations. However in rooms with greater volume, particularly in auditoriums, multipurpose rooms and gymnasiums, it is necessary to extend the phono-absorbent linings to the walls. If walls are available, it is necessary to use solutions with higher mechanical strength, which may be, for example, perforated panels and /or grooved

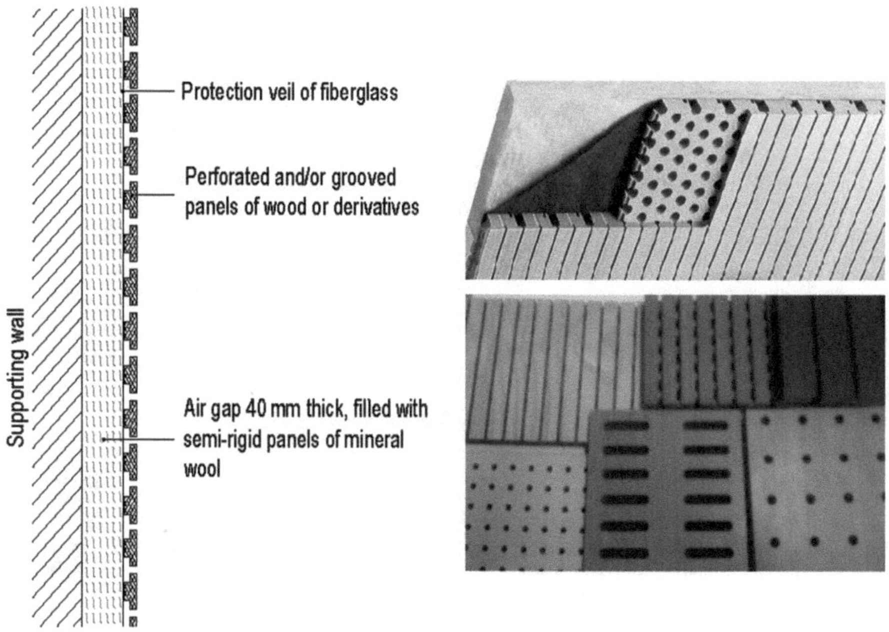

Fig. 2 Examples of constructive solutions used as absorbent linings in walls

panels of wood or derivatives, as illustrated in Fig. 2. Often there is some resistance to implementing phono-absorbent solutions in corridors, however, its application in ceilings is very important, even when classes begin and end at the same time, since it is a practice in some schools to teach with doors open or partially open. In these cases this solution minimizes indirect sound transmission among rooms.

The limitation of noise levels in a school, either produced by the occupants themselves, or generated by mechanical equipment, is one of the greatest difficulties in school buildings, particularly in buildings with limited constructive interventions. In relation to noise generated by occupants, actions hinged on the three aspects mentioned above usually allow solutions to these problems. As for the noise generated by equipment, in addition to acting in the three mentioned ways, it is still necessary to minimize transmission of vibration through the structure and the transmission of noise through ducts or directly to outside. For this purpose, inertia floating slabs (see Fig. 3) in the floor of the technical areas can be created, either outside or inside, which will support the equipment, fittings and pipes (which must be applied using flexible sleeves in the pipes, in the transition out of the technical area). In pieces of equipment that emit larger vibrations, application of anti-vibration supports in the equipment may also be justified.

The transmission of noise through the interior of ducts can be controlled through application of sound attenuators (see Fig. 4) between the HVAC equipment and ducts (in the extraction and/or in insuflaction). However, especially when the equipment is located within existing spaces limited in size, it is not always feasible to use these devices in ducts, and they also may promote the appearance of indoor problems related with air quality. It should be noted that the strong need to renew the air, imposed by RSECE, usually leads to application of high power sound equipment and of conducts with large sections, which when applied near the classrooms and other rooms sensitive to noise, can lead to values of LAr,nT substantially higher than those imposed by regulations. (Note that values up to 5 dB (A) above the limit do not usually generate frequent complaints, but above this value the problem can be severe.) The layout of the ducts can also cause breaks in the insulation of partitions that are run through them, which should be minimized, by choosing the most favourable routes and/or through the application of sound attenuators. With regards to transmission of noise to the outside, apart from possible problems of excessive noise inside the school, there may still be problems of noise annoyance to the closest neighbours. Minimization of this problem may involve the application of phono-absorbent barriers surrounding the equipment when applied in the exterior or by applying grids with sound attenuation, in the openings to the outside (see Fig. 4).

Typical Execution Problems

Besides the appropriate design of solutions and corresponding details, correct execution is crucial to guarantee a proper acoustic performance of buildings. It is common to find the same kind of constructive solutions applied in apparently similar situations, but with a completely different acoustic performance. The differences may be due to flanking sound

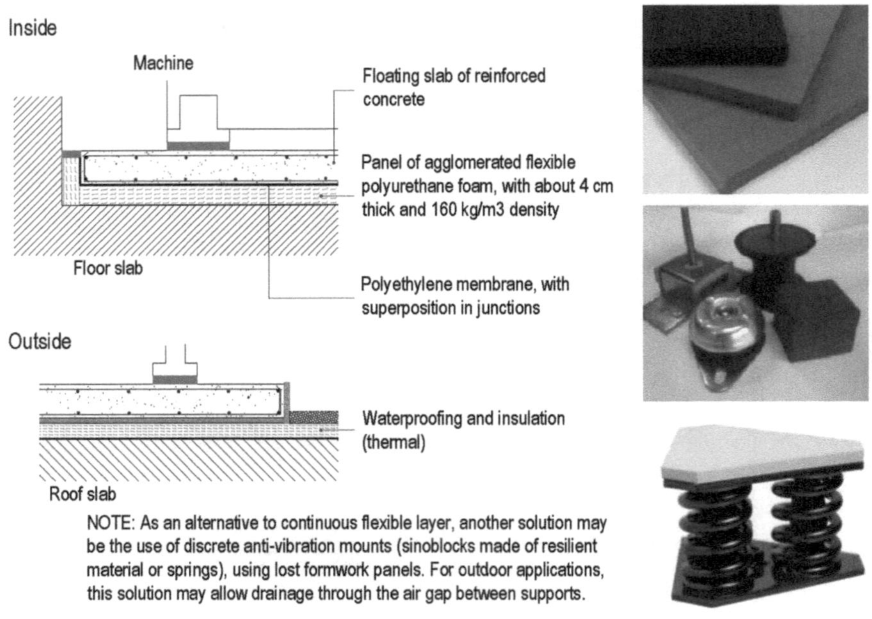

Fig. 3 Example of a constructive solution proposed for an inertia floating slab with a resilient continuous layer or with sonoblocks

Fig. 4 Images of constructive solutions used to interpose in the ducts (attenuators) in the exterior (acoustic barrier) and aperture into the exterior (acoustic grids)

transmissions, depending on how the separation elements are interconnected and their specific characteristics, but, quite often, large differences are found due to the construction process and to decisions made during construction. As in other areas but with particular relevance in sound insulation, achieving success in construction requires a thorough knowledge of the materials used and of the construction technologies applied. A constructive solution with a predictable high acoustic performance can result in complete failure, if some mistakes in its execution are committed, even of very small size, which are usually unnoticed during construction. It should also be noted that, in addition to the construction process, the link between the acoustic and the other specialties involved in the project, including architecture, stability and technical installations is fundamental.

In airborne sound insulation, either in case of insulation between the exterior and interior (facade insulation), or in the case of insulation between rooms, typical problems that can greatly compromise the final results, are usually due to the existence of leaks (commonly in the glazing frames and possibly in blind boxes) and/or due to the existence of "weak points of insulation," such as: the masonry joints which are frequently poorly filled with mortar; the joint between the last row of bricks and the lower face of the slab is barely topped,

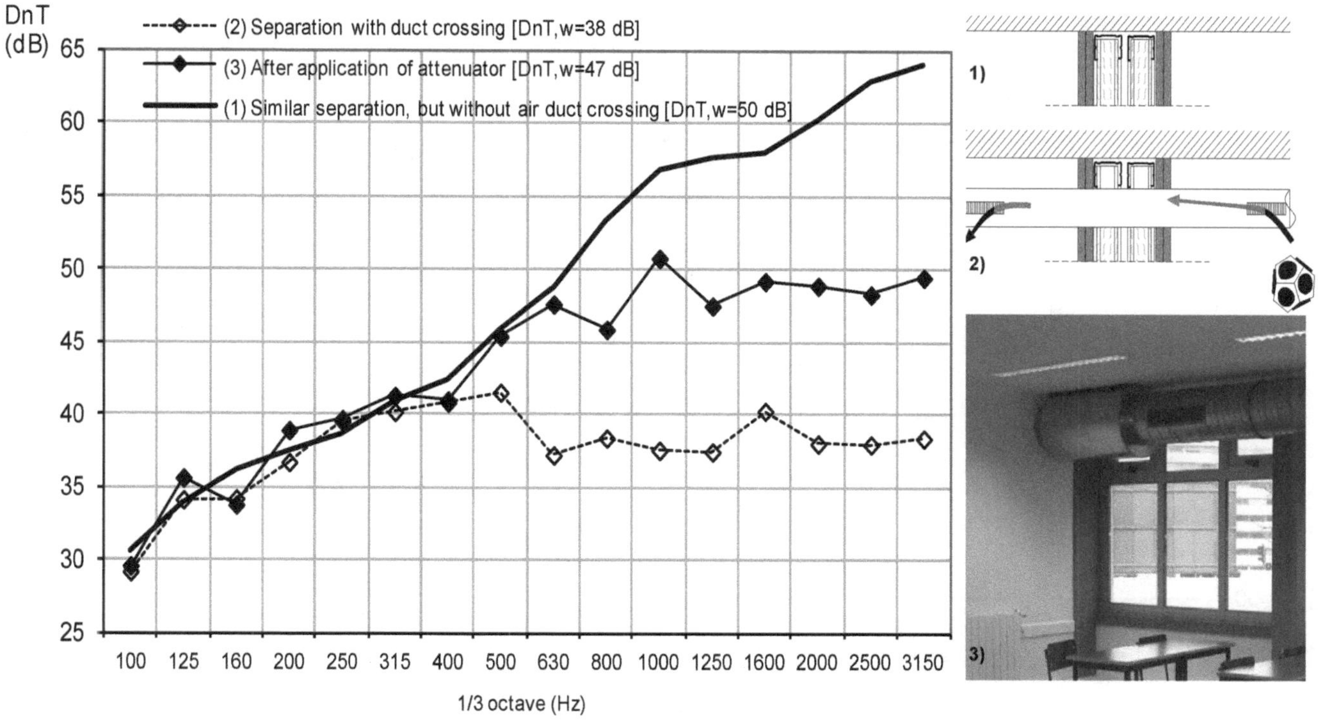

Fig. 5 Airborne sound insulation between adjacent rooms separated by a double gypsum plasterboard partition with and without duct crossing

often exacerbated by the small thickness of the lining of the walls; the boxes embedded in separation walls (mainly in the situations of lightweight partitions, where the plug/box breaks throughout all the thickness of the wall panels) and the crossing of ducts and/or courettes. Figure 5 shows, for example, results of sound insulation between adjacent classrooms for three cases: with ducts directly crossing two adjacent rooms, without this crossing and with the ducts crossing the wall but with a sound attenuator placed just before the crossing.

In the impact sound insulation, where often floating screeds are used under the floor covering, or also vinyl or linoleum with flexible base are used, problems arise mainly in the first case. Indeed, in most situations, the performance of the floating screed is very weak, owing to mistakes in construction, often of very small size, as is the case of rigid connections created by the adhesive used to settle the floor covering, especially when the covering is of ceramic tile or stone (see Fig. 6). Moreover, when opting for lightweight partitions (gypsum plasterboard), sometimes the construction begins by executing a continuous floating screed between adjacent rooms, which increases strongly the impact sound transmission, often resulting in a worse output than that without the floating screed (see Fig. 7).

Another problem that often arises from school buildings is excessive noise, originated in mechanical equipments, transmitted into the interior of the school and/or to the surroundings, sometimes exacerbated due to wrong decisions taken during construction and because of the lack of compatibility between different design specialities. Examples of these situations are non-application of sound attenuators, usually justified by the lack of available space, excessive speeds of airflow in ducts (sometimes with flow reductions of 25 % it is possible to achieve reductions in noise level above 5 dB (A)), the option for higher power sound equipment than those defined in the design, and changes in the position of the equipment that often occurs during execution, sometimes as an attempt to solve design omissions.

Although the displayed results correspond only to case studies, they allow us to demonstrate that a successful construction in the acoustic point of view is very dependent on the construction process. The existence of construction defects, even in very small dimension, such as some wrong decisions taken either during the design stage or during construction, can lead to very weak results.

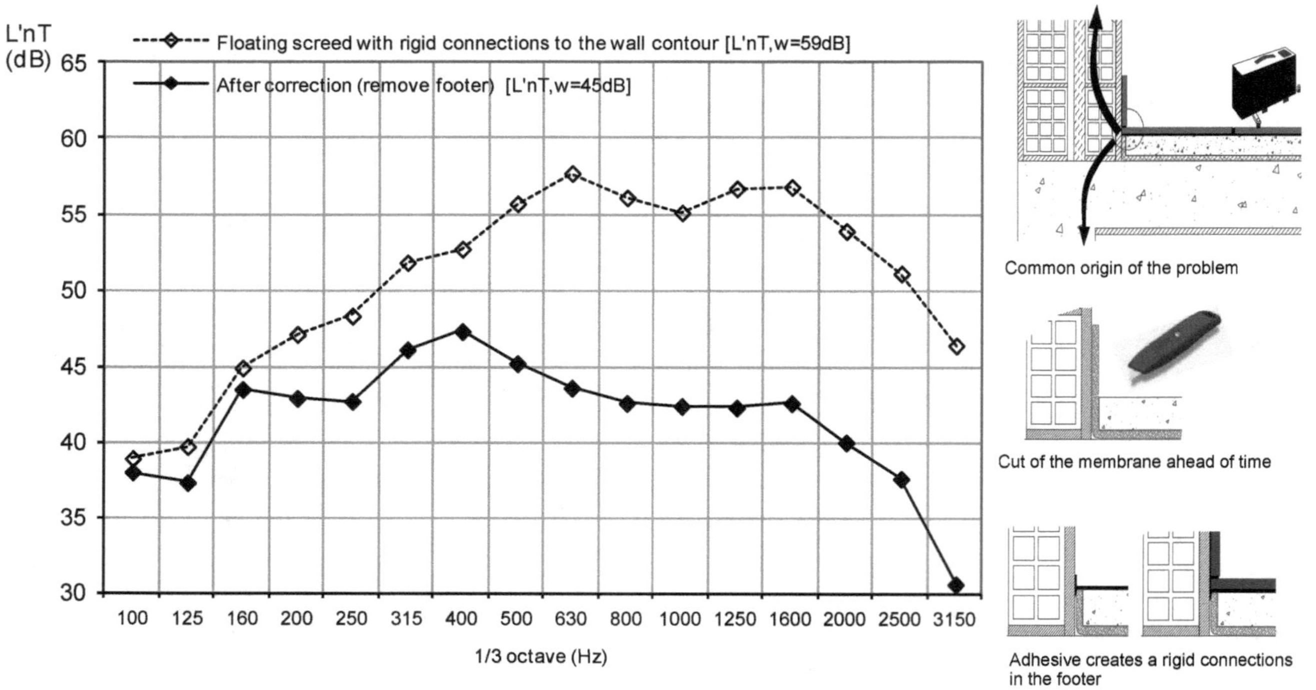

Fig. 6 Measured padronized impact sound pressure level, with sound transmission from the lower (sound generated has a strong impact sound component) to the upper level (receiving space to protect from sound), with rigid connection in the footer and after pulling off the footer around the entire floor

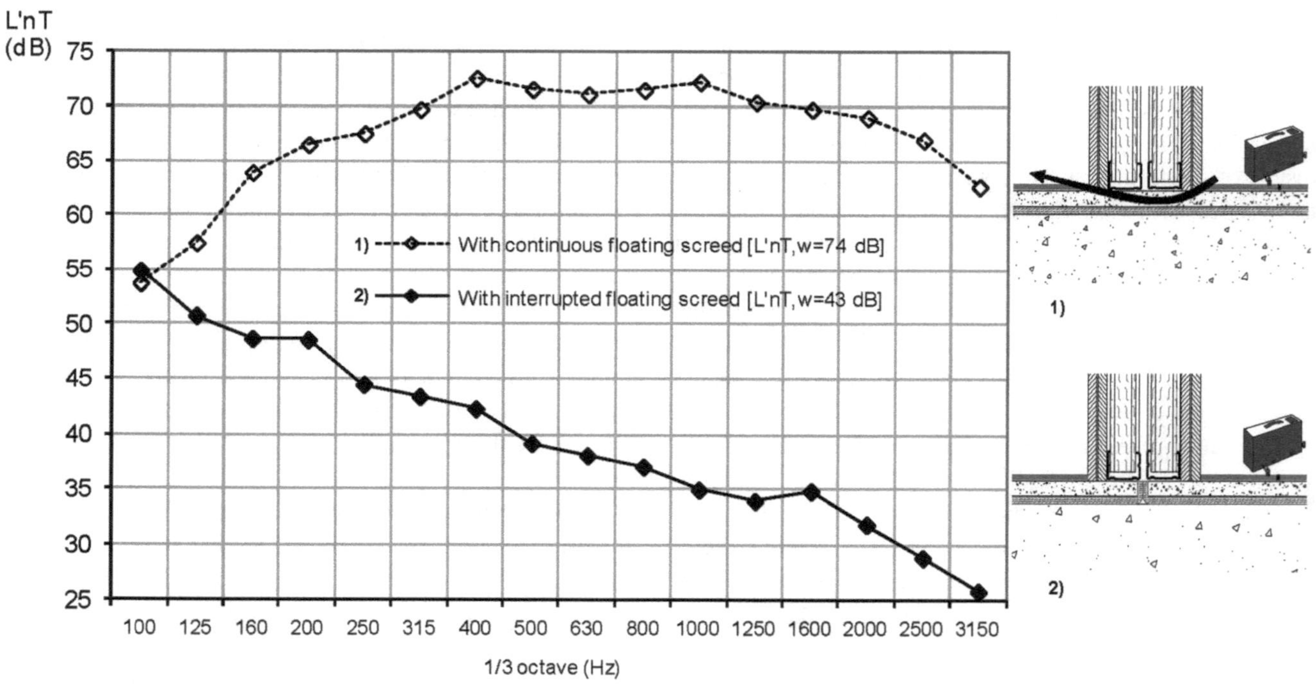

Fig. 7 Padronized impact sound pressure level between rooms of the same floor with and without continuity in the floating screed

References

1. RRAE – Portuguese regulation on acoustic requirements for buildings, approved by Decree Law n° 129/2002 of 11/05 with new wording given in Decree Law n° 96/2008 of 09/06
2. ANSI S12.60-2002 (2002) Acoustical performance criteria, design requirements and guidelines for schools. American National Standards Institute, Washington, DC
3. RGR – General Portuguese regulation on noise, approved by Decree Law n° 9/2007 de 17/01

Maintenance and Maintainability of University Buildings the Challenge of Façades

Rui Calejo

Building Management

Definitions

"*Latus sensus*", Management has the purpose of optimizing the activities of organizations by taking decisions on the options to be taken in a pondered and conscious way. Such decisions should always be well grounded, based on a substantial amount of data and should promote the fulfilment of the goals it undertook. In face of what has been said, we can thus consider that Maintenance fits into a wider area known as Building Management.

It is thus considered as relevant to define three different areas within the scope of activities of University Buildings Management: technical, social and economic.

Economic Management

In order to understand the complexity of economic management one needs to understand the concept of a building global cost. It largely overcomes the initial investment at the time of its acquisition. Being a real estate asset that has to be preserved, deferred costs, during the service stage of the building, take up a substantial share of such global cost.

Maintenance and usage costs represent about 80 % of the building global cost. [1]

Let us analyse Table 1.

As can be understood, this type of management is much affected by the typology of the building and by the level of quality required from it, as well as the Maintenance Strategy.

R. Calejo (✉)
Faculty of Engineer, Oporto University, Porto, Portugal
e-mail: calejo@fe.up.pt

One of the methods used for calculating the building global cost is the Life Cycle Cost (LCC).

$$CG = CI + \sum_{n=1}^{n=N} \frac{Cam + Cae.Cau}{(1+a)^n} + \sum_{k=1}^{k=N/M} \frac{Ccm}{(1+a)^{kM}}$$

In which:

CG represents the Global Cost in Euros;
CI represents the Initial Cost in Euros;
Cam represent the annual maintenance Costs in Euros;
Cae represent the annual operating Costs in Euros;
Cau represent the annual usage Costs in Euros;
Ccm represent the cyclic maintenance Costs in Euros;
M represents the Periodicity of cyclic costs per year;
N represents the Service Life in number of years;
a represents the equivalent average annual discount Rate.

Functional Management

Considering that university buildings require different measures, functional management takes up specific characteristics in this class of buildings. This activity does not include technical aspects although it does frame them. At this level management is responsible for hiring third parties in order to intervene in the building, for establishing and applying rules of use which are fundamental for university buildings:

- Permission and access to the different areas
- Identifying users
- Movement of vehicles
- Loading and unloading

The concept of Maintainability has full application at this level as the building should facilitate this social aspect of its maintenance through its built-in solutions.

V. Abrantes et al. (eds.), *The Pre-Fabrication of Building Facades*, Building Research: Design, Construction and Technologies, DOI 10.1007/978-3-319-22695-8_10

Table 1 Total cost of a building

LCC			
Initial Costs	**Differed Costs**		
	Exploration	**Maintenance**	**Use**
Land	Energy	Inspection	Staff
Design	Management	Cleaning	Materials
Construction	Water	Pró-action	Taxes
Licences	Sewage	Correction	
		Replacement	

Technical Management

Includes all interventions for the proper functioning of the building solutions, namely regarding maintenance actions. It is concerned with the building's performance during its service period, in other words, during the period of time in which the building at least preserves its minimal performance requirements.

Technical management can be divided into six fundamental areas such as maintenance, cleanliness and hygiene, emergencies, safety, functional adjustment and legal compliance. [2]

A technical manager must master the existing maintenance policies in order to be able to act effectively. As far as cleanliness and hygiene, a clear distinction must be made between general cleanliness for cleansing common areas of the building, which should be performed very often, and technical cleansing. As far as technical and accidental emergencies, the manager's role might include solving such issues or to the simple task of calling over someone to solve them. As far as safety, the role of domotic is becoming more and more present. Any functional disadjustments and the initiatives to solve them are the responsibility of technical managers, as well as complying with legal requirements.

Domotic and Maintainability in Technical Management

Maintainability is much benefited whenever we are dealing with "Smart Buildings", in other words, buildings endowed with automated processes. IT and technological developments enabled automatic management of certain aspects such as energy management, cleanliness and safety. Such IT systems work through a database which has been collected in an automatic way by such systems, then recorded and analysed in a comparative way against standard values.

Presently, smart buildings can bring about differences in terms of safety, gate control, central control, video surveillance, project management, waste reduction or even in terms of comfort. As a matter of fact, smart buildings can generate interesting discussions. Although, on one hand, its implementation is conditioned by economic factors related to the buildings, the gains introduced by this system as far as minimizing resource waste can minimize the costs related to its implementation in a given building. In fact, investment costs with automation correspond to about 10 % of the total construction cost, and the return is about 30 % of the water, electricity and gas expenses. [3]

A smart system should be able to integrate a system, to be executed in the most diverse conditions, to be reprogrammable in an accessible way, should be understandable by the user, should have enough memory and self-correction capabilities.

In the future, smart buildings can take up a greater and greater role in terms of building Maintainability, as practically all these initiatives can be facilitated through IT systems. Namely, for instance, air exhaust systems may become automatic and integrated into a domotic system for such simple processes such as automatic opening and closure of air pathways ways, thus preventing improper use of these elements, ensuring that minor mistakes are not committed whenever the system is used. Another option would be, for instance, automation of fuel doses in a pellet boiler.

At European level, university buildings are major energy consumers, representing a share of about 30 % of energy consumption by all types of buildings. [4]

Taking advantage of the smart building concepts, in the 1980s, the concept of centralized technical management emerged. Its main areas of intervention are electrical facilities and mechanical systems, comfort, energy management, protection, maintenance and safety. "In a study carried out by Ecofys and Fraunhofer ISI, which was recently presented in Brusells, it is shown how just in university buildings the EU can save up to 23 billion Euros per year in 2020 if it adopts the tightest efficient energy management targets." [5]

Costs Related to the Lack of Maintainability

What has been said so far highlights some of the points in which lack of Maintainability generates costs. According to Mirshawka and Olmedo [6], the costs generated by lack of Maintainability are just the tip of an iceberg. This visible tip corresponds to the costs with man-power, tools and instruments, materials applied during repairs, outsourcing costs and other costs related to the facilities occupied by the maintenance team. Below this visible tip of the iceberg are the major invisible costs, following from unavailability of decisive equipment for University buildings.

Unavailability costs are mostly focused on those following from the production losses, poor product quality, production recomposition and commercial penalties, with possible consequences upon the company's image. These same aspects were also dealt with by Cattini [7], when he points out the costs connected to equipment unavailability and deterioration as a consequence of lack of maintenance. This relationship between maintenance costs, unavailability costs and productivity was analysed by Chiu and Huang [8], with a mathematical model which conclusion points towards a better relation cost-benefit whenever maintenance is treated preventively, instead of uncontrolled situations of the productive process due to lack of maintenance. Considering maintenance as a premise for bringing down production costs, the best policy for optimizing costs should be defined and adopted. This analysis can be observed in the classical diagram shown in Fig. 1, illustrating the relationship between preventive maintenance costs and the cost associated to failures.

The main costs associated to failure are basically the parts and manpower required for the repair, plus, mostly the unavailability of the equipment. The diagram in Fig. 1 shows that growing investment in preventive maintenance reduces costs associated to failures, and consequently decreases the total maintenance costs, in which the costs with preventive maintenance are added to failure costs. Meanwhile the diagram also shows that, beyond the optimal point as far as investment for preventive maintenance, more investment brings about little benefits for the reduction of failure costs and ends up increasing the total cost.

This issue was analysed by Murty and Naikan [9] who work on the limits of availability and present a mathematical model for calculating the optimal availability point, as shown in the diagram of Fig. 2.

Maintainability Handbook

There should be a service handbook for new buildings, in their project stage, or for existing buildings, in their service stage, whose rehabilitation has been, at least 50 % of a similar new construction. On one hand, the purpose of this document is to supply information to users or to expert technicians on how to easily and properly carry out the maintenance of a given component, and, on the other hand, to teach the user how to use that same component. In other words, ultimately it is divided into two domains which shall generate two different documents. The "Maintenance Handbook" and the "Service Handbook".

The "Maintenance Handbook" explains in detail the maintenance management sector, by describing the procedures and interactions between processes. Inspection routines for each component are explained in this handbook, as well as intervention strategies for each state of performance. This document should also include a list of elements with maintenance requirements in each building, an inspection form describing which interventions have already been performed, the different materials and a description of the building, a form describing which anomalies have been found and an inspection report briefly summarizing the inspection that has been carried out (Table 2).

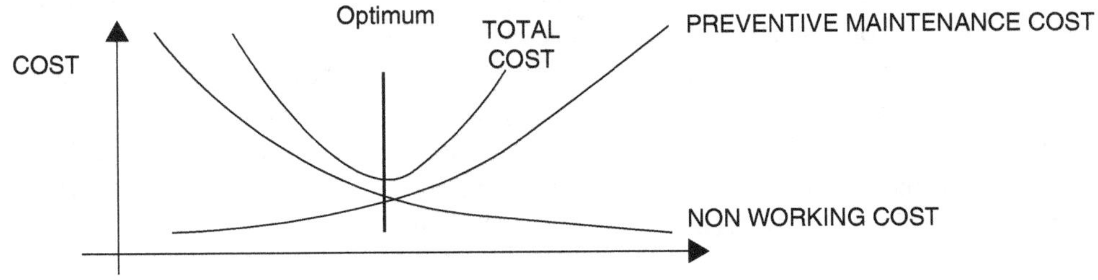

Fig. 1 Diagram with costs versus level of maintenance [7]

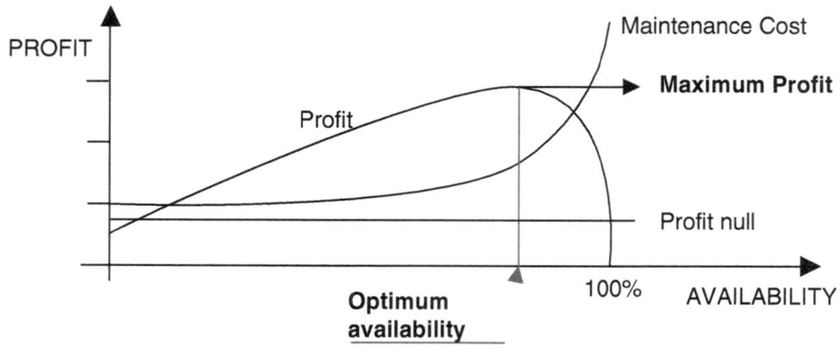

Fig. 2 Diagram of profit versus availability [9]

Table 2 Organization of a service handbook

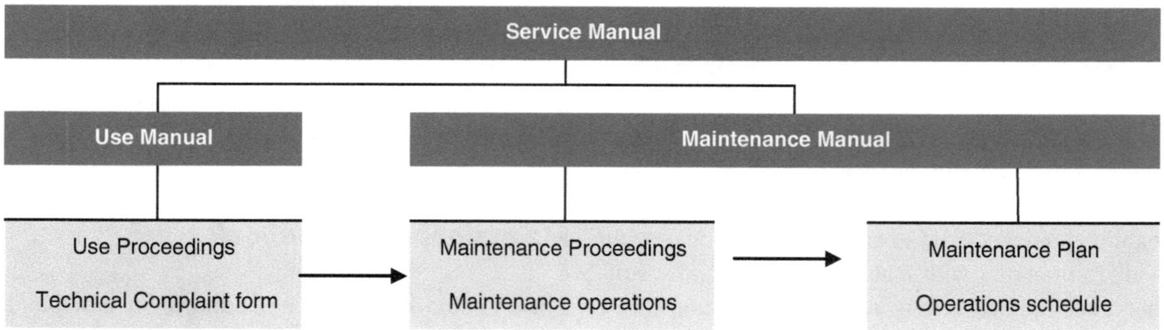

The "Service Handbook" is specific for the building's type of activity and should include the following: [10]:

Characteristics of the different components and elements of the building;
References of the materials that have been applied and the equipments that have been installed;
Guarantees and information on the suppliers of the different components
Users' rights and obligations;
Notices for appropriate use;
Indications of the material to have available at all times, for any eventual repairs;
Legislation and regulations related to housing units;
Actions to be carried out when entering into the building.

Conclusions and Proposal of Developments

The concept of maintainability in university buildings has certain specificities that set them apart from all other buildings. The costs generated by non-maintainability as well as those following from the absence of pre-established procedures – a handbook – have been detailed in this text.

In university buildings, the role of building managers shall have an increased weight. At present it is difficult to draw conclusions from the mistakes and the consequent costs in the past in the field of construction as very few buildings have information on the interventions carried out in them. It is necessary to systematize information in order to gather knowledge that may generate standard interventions to be made. To set up a standard database can be the answer to such issue. Following technologic developments, computer *software* is becoming more and more optimized for building management.

As far as future perspectives, bar codes and *Radio Frequency Identification* (RFID) are both methodologies to help control maintenance to the components of a building throughout its service life cycle and which shall perform an important role in the future of maintainability.

A good way to standardize and inspect the best way of performing maintenance to a given component may have to do with the existence of a reference of such component, with its basic data. The two main benefits from these two approaches are the building's installation date, how regular interventions need to be carried out, and what is the proper way to perform such interventions and by whom.

It is well known that most building users/managers are not aware that all of its components undergo degradation with time and need regular maintenance. The existence of a building handbook makes information available to users, but has a major shortcoming: users must resort to it in order to recall or to even know that they must carry out maintenance actions. The existence of a bar code or of a RFID system may make up for such shortcoming.

Acknowledgment The author wishes to thank to Engineers Ruben and Gabriel Pinto Teixeira for their contributions to the basic text as well as the production of images.

References

1. Perret J (1995) Guide de La Maintenance des Bâtiments. Paris, Moniteur Référence Technique, França
2. Calejo R. (2001) Gestão de Edifícios – Modelo de Simulação Técnico-económica. Dissertação de Doutoramento em Engenharia Civil, Faculdade de Engenharia da Universidade do Porto
3. Bolzani C (2009) Residências Inteligentes. Editora Livraria da Física, Brasil
4. Carpinteiro J (2009) Aquecimento, Ventilação e Ar Condicionado. Verlag Dashofer, Lisboa
5. http://www.climatizacao.pt/noticias/21-aabril-b.aspx. Accessed on April 2011
6. Mirshawka V, Olmedo NC (1993) Manutenção – combate aos custos na não-eficácia – a vez do Brasil. Editora, São Paulo
7. Cattini O (1992) *Derrubando os Mitos da Manutenção.* STS Publicações e Serviços, São Paulo
8. Chiu H-N, Huang BS (1996) The economic design of x control charts under a preventive maintenance policy. Int J Qual Reliab Manag Camb 13(1):61–71
9. Murty ASR, Naikan VNA (1995) Availability and maintenance cost optimization of a production plant. Int J Qual Reliab Manag Camb 8(1):7–39
10. Calejo R (2006) Manuais de Manutenção e Utilização. Encontro Nacional sobre Qualidade e Inovação na Construção, LNEC, pp. 465–476, LNEC Edições, Lisboa

FSC
www.fsc.org
MIX
Papier | Fördert
gute Waldnutzung
FSC® C083411

Kolibri 360 GmbH
Ferdinand-Jühlke-Straße 7
99095 Erfurt, Deutschland
produktsicherheit@kolibri360.de